ARCHITECTS WHO HAVE WORKED WITH COSENTINI ASSOCIATES, 1952 - 2012

Brown & Associates • **E** • Edelman Sultan Knox Wood Architects • EDI - Houston • E
Ehrenkrantz Eckstut • Ehrenkrantz Eckstut & Whitelaw • Ehrenkrantz Group • Einh
Eisenman Architects • Eitington & Schlossberg • 1100 Architect • Elkus Manfredi Archite
Englert • Environetics Architect • The Environments Group • Ethelind Coblin Architects •
Collaborative • JJ Falk Design • Faridy Veisz Fraytak • Farinella and Sam Architects • Fau
Feher Architect P.C.• Roger Ferris + Partners • FGM Architects • Michael Fieldman & Partners • Fifield Piaker & Associates • Finn & Jenter • Karl Fischer Architect • Fisher & Spillman Architects • Flad & Associates • Fletcher Thompson • Foster + Partners • Foster Wheeler • Fox & Fowle Architects • Herbert Fox Associates • Fradkin & McAlpin Associates • Francis Cauffman Foley Hoffmann • Franke Gottsegen Cox • Frank & Marcotullio Design Associates • Ulrich Franzen & Associates • Fredenburgh Architects • William Freed • Freeman & Pizer Architects • Freidin Bolcek Associates • Freidin Kleiman Architects • Freidin Studley Associates • Friedmutter Group • Glen Fries Associates • Arthur Froelich & Associates • Frederick Frost & Associates • Frederick G. Frost • William J. Fryer • Joe Fuchida • Fujikawa Johnson & Associates • Fujikawa Conterato Lohan • Fujikawa Johnson Gobel Architects • Fuller & D'Angelo • Fulmer & Bowers • Furnstahl & Simon Architects • FXFOWLE Architects • **G** • Michael Garvin • Gatje Papachristou Smith • Alan Gaynor & Co. • GBQC Architects • Geddes Brecher Qualls Cunningham • Gehry Partners • Genovese & Maddalene • Gensler • Gerner Kronick + Valcarcel, Architects • Irving Gershon & Jack Brown • Gertler & Wente Architects • Gertler Wente Kerbeykian • Gibbons & Heidtman • Gibbs & Cox • Sidney Philip Gilbert Associates • Gilbert Architects • Gillis Architects • Gilliemo Colado • Gilsanz Murray Steficek • Francis X. Gina & Associates • Robert Ginsberg • GJB Design Associates • GK Studio Associates • Gluckman Mayner Architects• Richard Gluckman • GMW Partnership • GNA Architects • Goettsch Partners • Raymond Gomez & Associates • Teodoro González de León • Martin Goodman • Goody Clancy & Associates • Gordon & Levin • Gordon Bajbek Barren • Jack L. Gordon • Gordon Beard Grimes & Bahls • Peter Gorman • Goshow Associates • Gould Evans Associates • Frank Grad & Sons • Grad Associates • Grad Partnership • Adrian Grad Design • Graham Gund • Granary Associates • Office of Jacques Grange, New York • Michael Graves & Associates • Michael Graves Architect • Robert Gray Architects • Allan Greenberg • Greenberg Farrow • Greenwell Goetz Associates • Gregg & Wies Architects • Gresham Smith & Partners • Griswold Heckel & Kelly Associates • Gronske Design • Amie Gross Architect • Harold Charles Gross Architect • Group 3 Design • Gruzen Associates • Gruzen & Partners • Gruzen Samton Architects • Gruzen Samton Steinglass • GSB Group • Guenther 5 Architects • Guenther Petrarca • Gund Partnership • Gwathmey Siegel & Associates • **H** • H2L2 Architects • H3 Hardy Collaboration Architecture • Zaha Hadid Architects • HAEAHN Architcture • Hague Richards • Haines Lundberg Waehler • HakSikSon, FAIA • Peter Halfon Architect • Lawrence Halprin & Associates • Hammond Beeby & Babka • Handel Architects • Gary Edward Handel & Associates • John Hanna Architect • Hanna/Olin • F.T.L. Happold • James Harb Architects • Hardy Holzman Pfeiffer Associates • John Hardy • Harman Jablin Architects • Alonzo J. Harriman Architects • Harrison & Abramovitz • Harrison, Abramovitz & Abbe • Harrison, Fouilhoux & Abramovitz, Architects • Hartman-Cox • Hart Associates • Richard Hayden • Heerim Architects • Heerim/Kunwon • The Heiserman Group• Helbing Lipp • Hellmuth, Obata + Kassabaum • Hellyer Schneider & Company • David Helpern & Associates • Helpern Architects • Henningson Durham & Richardson • Herzog & de Meuron • HGA/KKE • Hickok Warner Fox • Hillier Architecture • HKS • HLW International • Hnedak Bobo Group • HNTB • Holabird & Root • Steven Holl Architects • Horowitz & Chun • Stanley L. Horowitz • HSS • HTI International • David Alfred Hunter Architects • Huntsman Architectural Group • Huygens, DiMella, Shaffer & Associates • **I** • IA Interior Architects Boston • ICON Architecture • Ikon 5 Architects • Ingram-Sageser • Integrated Design Group • Interior Architects Collaborative • Interior Space International • INW • I.S.D. • Arata Isozaki • Iu + Bibliowicz Architects • The Ives Group • Ives Turano & Gardner • **J** • J&K Interiors • Stephen B. Jacobs • David Jacobson Associates • Steven G. Jacobson • Jacquemin Beringer Architects • JBA Interiors • JCS Design Associates • The Jerde Partnership • Jessop Architects • Jeter Cook & Jepson • JINA Architects • Gerald Johnson • Johnson/Burgee Architects • Philip Johnson • Philip Johnson / Alan Ritchie Architects • Johnson-Wanzenberg Associates • Johnson Jones Architects • Johnson Jones/Gruzen Samton • Jones Lang Wootton • Jones Lang Lasalle • Wendy Evans Joseph • JPC Architects • JRS Architect • Jung/Brannen Associates • Juniper Russell & Associates • Juster Pope & Associates • **K** • KA Architects • Kaeyer Parker & Garment • Stephen J. Kagel Associates • Albert Kahn Associates • Louis I. Kahn • Kahn & Jacobs • Kajima International • Kallmann & McKinnell • Kallmann, McKinnell & Wood Architects • Kanan, Corbin & Schupak • Kaneko Architects • Richard Kaplan Architect • Karahan Schwarting Architects • Karco Davis • Karlsberger Architects • Kobi Karp Architecture • Karplus & Nussbaum, Architects • Kaufman Architecture • Gene Kaufman, Architect • KBJ Architects • Kellenyi Johnson Wagner • Kellner Plofker Architects • Kelly & Gruzen • Kendall/Heaton • Kevin Kennon Architects • Keogh Design • Heath Kessler • Ira Kessler, Architect • Ketchum Gina & Sharp • Keyes, Condon Florance • Keyes, Lethbridge & Condon • Tai Soo Kim Partners • The Klein Company • Milo Kleinberg Design Associates • R.M. Kliment & Frances Halsband Architects • Kling • Kling Lindquist Partnership • Vincent Kling • Kling Stubbins • KM2 Architectural Studios •

continued on back inside cover

THE INVISIBLE ARCHITECT

ISBN: 978-0-9884962-0-0

Published in the United States by Piloti Press for
Cosentini Associates, a TetraTech Company

Printed in Canada

THE INVISIBLE ARCHITECT

Marvin Mass, P.E.

with Janet Adams Strong, Ph.D.

TO THE WORLD OF ARCHITECTS

This book is dedicated to the many architects I have worked with during the last 60+ years.

Like many of you, I started in the industry just after World War II. I built my career on the facilities that you designed and that helped make our country great.

I'd like to thank you for the friendship you have extended over the years and the knowledge you so generously shared. There is no question in my mind that my success as an engineer results from the wonderful opportunities that you and your buildings afforded.

With this book I hope to express my gratitude and appreciation.

I. M. Pei, Ruth Mass, Marvin, Philip Johnson

CONTENTS

Marvin Mass: The Invisible Architect

FOREWORD

As with all successful people, there are many aspects of Marvin's life that vie for attention: the professional, the family man, and personal enjoyment. Somehow he can combine all three of these in one, approaching everybody in the same open, congenial manner. In a profession whose practitioners are caricatured as interested only in machines or systems of operation, Marvin puts people first. He makes no distinction within the great garden of personalities that he meets. Whether clients, friends, business associates, or family, Marvin receives them all with the same sanguine spirit of welcome.

When he explains that buildings, no matter how laden with technology and machinery, exist for the benefit of people, his great professional skill can make the built environment so pleasant and easy to maintain that the experience of architecture becomes a public delight, without the public even knowing why.

Through an ingenious investigation into the efficient use of energy, Marvin has also produced a body of work that includes many trend-setting achievements. Accomplished via the use of highly practical systems that often use reclaimed heat from lighting, computers, or sunshine to reduce the load on conventional boilers, he has been in the forefront of energy-efficient mechanical equipment and system design. In addition to refrigeration, he has also used natural forces for cooling.

Early in his career Marvin challenged manufacturers to provide machinery designed to offer small-scale areas of heating and cooling so that individual needs in small areas could be addressed without operating whole floors or entire buildings. This emphasis on how people actually use space, instead of relying upon abstract numbers that calculate occupancy but don't describe need, can offer great savings in the use of energy. By this very sensible approach each area receives the temperature and humidity it requires.

Marvin's integration of the entire professional design team in the conceptualizing of a new structure becomes part of the planning process. It has moved mechanical engineers from a peripheral role into the center of building design. As a direct result, their systems became as integral a part of the building as the floors, walls, and ceilings. This collaborative approach to design recognizes how a better understanding of program and architectural intent can directly benefit engineering.

Marvin has always had the ability to respond to an amazing variety of architectural talent with good grace, despite some of the unwieldy egos involved. It is testament to a remarkable man, one to whom generations of talented and sometimes irascible designers have responded with affection and admiration. It should be recognized that in continuation of his numerous accomplishments over six decades, Marvin still works in his office every day, as vital as ever. Seeing this compendium helps us launch his next decade, making visible a full presentation of remarkable success.

The many discoveries and achievements that follow will give you both wonder and delight.

Hugh Hardy, FAIA

ACKNOWLEDGMENTS

Beyond all of the architects who made this book possible—and necessary—I would like to thank the many people whose personal efforts were instrumental in making *The Invisible Architect* a reality.

I am especially grateful to my good friend Hugh Hardy, who generously wrote the Foreword, and also to David Childs, Bruce Fowle, Frank Gehry, Jill Lerner, Alan Ritchie, Kevin Roche, and Nancy Ruddy, each of whom kindly contributed their thoughts. It has been my privilege to know these wonderful people over the years and I am delighted they could take part in this celebration of my career.

In addition, I must be honest and say that this book would not have been written without the great help of Dr. Janet Adams Strong, a long time friend of architects and engineers. Without her there would be no book. Graphic designer Linda Zingg, Janet's partner at Piloti Press, did a terrific job in developing the layout. It was a genuine pleasure to work with these two dedicated professionals and I highly recommend them.

It is also my pleasure to acknowledge Charlotte Meyerson, who helped me several years ago to start recording my thoughts. Without such an important first step, this book might have remained just a dream.

Along the way I have, as always, relied on the good efforts of my unfailing assistant Kay Lawrence, and, of course, my wife Ruth, the proverbial woman behind the man, who urged me to document my life's work. I am proud that my son Doug contributed the final chapter to this book as he so ably leads Cosentini into the future.

For help in supplying illustrations, I am pleased to thank Frances Halsband for sharing her classic photo of the bespectacled tribute to Philip Johnson in 1993. Alan Ritchie and Aldo Cossutta also supplied images, as did the gifted photographer Alan Gilbert. I am also happy to thank Emma Cobb of Pei, Cobb Freed & Partners, Linda Scinto from Kevin Roche John Dinkeloo Associates, Jay Shockley of the New York City Landmarks Preservation Commission, and Kate Wood, executive director of Landmark West. Kristin Marshall of the State University of New York at Albany, and Stephanie Coleman of the Ryerson and Burnham Libraries at the Art Institute of Chicago were also very helpful.

Within Cosentini Associates, former Creative Manager Karen Norgren offered photo assistance as did our Administrative Manager Carol Rizzo. Gretchen Bank, Cosentini's new Director of Business Development, has been invaluable in finalizing the book. Communications Coordinator Cori Carl has also been a great help in securing photos and permissions.

If I have forgotten anyone, I apologize. I'll make sure to include you in my next book.

I hope that everyone who reads this book will appreciate the extraordinary developments that have taken place in architecture and engineering over the past half century. I want to give special thanks to our parent company, TetraTech, for its support of *The Invisible Architect.* TetraTech's confidence in inviting us to join them has allowed Cosentini to expand throughout the United States and around the world.

Marvin Mass with Kay Lawrence, his assistant of 45 years

At the annual gathering, friends and colleagues trade laughs and stories, and take stock of the year's events. 1984 brought the sad news that a dear colleague had died, only for him to arrive, quite alive, at the door. The matter was straightened out, and a legend was born in this letter of clarification.

Robert Wagenseil Jones & Associates Architects

106 Montowese Street Branford Ct. 06405

203 488 6283

16 January 1984

Mr. Marvin Mass
Cosentini Associates
2 Penn Plaza
New York, New York

Re: Determination of Life Status:
R. W. Jones

Dear Mr. Mass:

At the urgent request of Mr. Jones, I am writing as Medical Director of this large, prestigious and international architectural, planning, design and storm-door company in order to keep you current on the status of Mr. Jones...lifewise.

It is our firm's policy to subject all of our staff to life-status tests no more than every ten years in order to find out who is alive and/or dead. Often we are asked to do this more often by some of our more prestigious and international clients...usually within a few days of their receipt of an invoice for the prior month's work.

Recently we have been flooded with such requests in regard to Mr. Jones and I am pleased to be able to tell you that, after undergoing a series of exhaustive tests, Mr. Jones appears to, in fact, be alive; at least in my opinion. While not all of the tests were conclusive, one in particular was: when delt a hand of five-card stud (one-eyed Jacks wild) from a deck of pornographic Swedish playing cards, his pulse rose from 1 $\pm$ to well over 3+ and definite signs of low-grade mental activity were noted. My medical assistant, Zelda, who administered the test believes it was the Queen of Hearts with the dancing donkey that did the trick. Therefore, we are continuing to bill Mr. Jones's time at the usual hourly rate.

Should your Medical Director wish to borrow our cards... or Zelda...please have him give me a ring.

Very truly yours,

Melvin Strongfinger, MD

MS:z

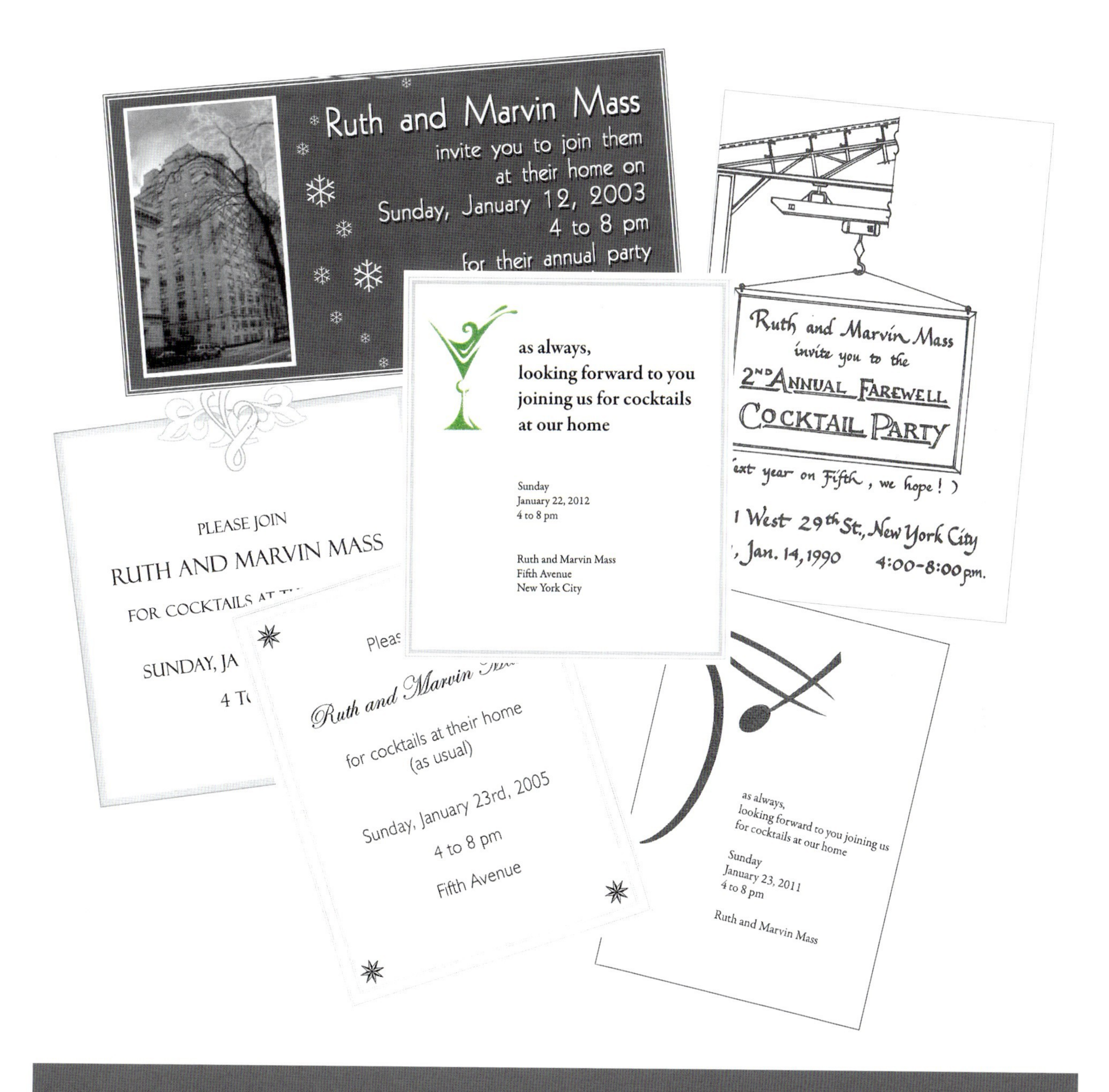

Every year for the past 42 years, Ruth and Marvin Mass have hosted a cocktail party for professional friends and colleagues from around the globe. The celebrated tradition was conceived to afford architects an opportunity to meet in a friendly, noncompetitive atmosphere.

Marvin A. Mass, PE

The Frank P. Brown Medal
awarded by
The Franklin Institute

April 12, 1989

In 1989, the Franklin Institute conferred on Marvin Mass its coveted *Frank P. Brown Medal*, the country's "foremost award honoring innovation and leadership in meritorious improvement in the building and allied industries."

Previous laureates include the architects R. Buckminster Fuller, Charles-Édouard Jeanneret (Le Corbusier), Louis I. Kahn, Pier Luigi Nervi, and Frank Lloyd Wright, as well as luminaries like Marie Curie, Thomas Edison, Albert Einstein, Henry Ford, Stephen Hawking, and Alfred Nobel, among others.

MARVIN A. MASS, P.E.

With the following presentation architect Melvin Brecher helped the international Committee on Science and the Arts to evaluate the degree to which the contributions of Marvin Mass "truly reflect the spirit, innovation and inspiration of [Benjamin] Franklin himself."

It is only fitting, on this 50th anniversary of the Frank P. Brown Medal, that engineer Marvin Mass be under consideration for this honor. Like Ben Franklin, he has given us practical applications for scientific discoveries and, as a result, has changed not only the way we live but also, the way we think about our environment. Franklin gave us the Franklin stove to heat our homes; Marvin Mass has given us innovative systems to heat and cool our 20th century structures. Franklin flew his kite and discovered that lightning is electricity; Marvin Mass showed us how to use and conserve energy—energy from the sun, from water, air, and, especially, from electricity.

In his 40 years as a professional engineer—the last 35 years as a partner and now senior partner at Cosentini Associates, one of the country's most prolific, innovative, and eminent mechanical, electrical, and sanitary engineering firms—Marvin Mass has transformed engineering practice and practices. He has, in fact, substantially changed the way our commercial and high-technology buildings are conceived, constructed, and maintained. His achievements in three principal areas—modular climate control systems, cooling technology, and re-use of waste heat—have set the standard in HVAC engineering and in the industry.

His approach is quite simple. He explains to his last-year architectural students at Harvard's Graduate School of Design, where he has taught for eight years, that "A building has more than a skin and bones; it also has a heart, veins, and nerves. They must all function together." In typically few words, he has summed up the dignity and relevance of his profession.

A modest man, Mr. Mass is well-known and treated with respect and gratitude by architects and building owners alike. But he's not famous. While considered to be one of this century's truly creative systems engineers, Mr. Mass holds no patents. Although he has caused an inordinate number of advances, he has not written or spoken extensively, even though he is an inveterate teacher. His few public appearances are for philanthropic reasons, proof of the respect and broad friendships he has earned. In fact, because his confident, visionary engineering has, in retrospect, a touch of the inevitable, many of his breakthroughs are taken for granted.

Perhaps because Marvin Mass has gone about his work quietly, few of us knew that, when we entered our offices this summer—one of the most uncomfortably hot summers on record—and turned on our air conditioners to the level we needed, it was Marvin Mass who had devised many of the systems that made our comfort possible.

ENERGY-EFFICIENT, MODULAR, CLIMATE-CONTROL SYSTEMS

Mr. Mass was the first to engineer energy-efficient environmental systems for high-rise office buildings based on the use of modular, climate-control units. Previous high-rise engineering technologies that existed in the 1950s and 1960s were total-building systems based on air-conditioning equipment designed not by professional engineers involved in the building industry, but rather by manufacturers. Cumbersome, complicated, and expensive, the equipment wasted energy because it was operated continuously at maximum capacity to heat or cool entire buildings to the level of greatest need. Working directly with the manufacturers, as is his preference, Mr. Mass encouraged the development of small, factory-assembled, self-contained, non-central refrigeration equipment units that cost considerably less to install, run, and maintain.

These modular units had any number of bonuses. For one thing, they enabled buildings to be sub-metered, which gave the user control over his own energy costs. For another, this singular innovation also dramatically affected how buildings could be physically organized. Occupancies could be organized according to need: hours of operation, adjacencies, and so forth.

Buildings could be programmed not just for space requirements, special uses, and comfort, but for energy efficiency as well. In the building industry, it is hard now to recall a time when it wasn't this way.

COOLING TECHNOLOGY: ENERGY FROM WATER AND AIR

In a career punctuated by "firsts," Marvin Mass's breakthroughs in cooling technology present a number of particularly outstanding achievements. Consistently relied upon to design the mechanical and electrical systems for a series of "World's Tallest Buildings," Mr. Mass harnessed energy from water and air to create innovative cooling methods. Many patented systems have their origins in his quest for answers to technological problems, and they have become standard practices. As Sir Francis Darwin summed it up (in 1914), "The credit goes to the man who convinces the world, not to the man to whom the idea first occurs."

First chilled-water storage system (1980): At 101 Park Avenue in New York City, in a building designed by Eli Attia & Associates, Mr. Mass created the first major chilled-water storage stem for office-building use in the country. It permitted a reduction of approximately 700 tons of equipment capacity in peak cooling load, which resulted in considerable cost savings. The system allows 500,000 gallons of water to be cooled at night, when energy rates decrease, and then circulated.

First use of an "ice-storage pond" in a commercial building (1982): For the headquarters of the Alabama Power Company in Birmingham, designed by Geddes Brecher Qualls Cunningham: Architects in association with the local office of Gresham, Smith & Partners, Mr. Mass took an idea from the milk industry and applied it to office buildings. Ice made on brine plates is flaked off into a storage tank below. As the ice gradually melts, it is used to cool the building. At night, the melt-off is sprayed back atop the plates to refreeze, preparatory to repeating the cycle.

First use of condenser watering cooling coils (1979): For the commercial office building located at 499 Park Avenue in New York City designed by I. M. Pei & Partners, Mr. Mass advocated the use of cooling tower water in pre-cooling coils. This system is engineered to take advantage of natural cooling that occurs as water evaporates. The evaporative process is used as a first source for cooling, with compressors—the traditional cooling mode—used only as a secondary, back-up source, thus considerably reducing energy loads.

To achieve this, Mr. Mass again worked with air-conditioning equipment manufacturers to develop a single piece of equipment that would make maximum use of available natural cooling to minimize operating energy consumption. The unit that resulted from this collaboration is capable of extracting cooling directly from a cooling tower water system to air-condition a structure, with minimum use of mechanical refrigeration.

A "building that God would heat and cool" (1977): One of the greatest challenges put to Mr. Mass, these instructions came from an equally unusual client, the Crystal Cathedral in Garden Grove, California, which had commissioned Philip Johnson/John Burgee Architects to design an 86,000-sq. ft. cathedral to seat 2500 worshippers. In order to cool this remarkable, virtually all-glass, 28-foot high space by natural ventilation, Mr. Mass advocated silver-coated reflective glass to reject the sum's heat. Not 10 percent of the sunlight shines through, yet at noon on a hazy fall day, the illumination is a bright as 500 foot-candles. As the sun moves across the sky, operable windows on all sides of the cathedral open to admit the natural drafts that create cross-ventilation. The "heart" of the system is a computer that logs and reacts to wind and sun conditions, and opens the windows spontaneously. This careful integration of mechanical systems has translated into an extremely low energy consumption.

RE-USE OF WASTE HEAT: ENERGY FROM EXCESS ENERGY

Always putting theory into practice, Mr. Mass says, "My whole life has been dedicated to real-life applications." For instance, 25 years ago buildings without boilers were unimaginable, especially in the dead of winter. Yet, as in this most amazing example of technical innovation, he was first to create environmentally comfortable spaces by using waste energy to heat, cool, and dehumidify. Interestingly, to make such reclamation systems work, Mr. Mass amplified and adapted cooling technologies.

Storing excess heat (1981): The AT&T Long Lines Building in Oakton, Virginia, is a 400,000-sq. ft. office building designed by architects Kohn Pedersen Fox Associates. Mr. Mass designed one of the first total heat-reclamation systems that uses refrigeration compressors to extract heat—from interior lights, people, the sun, and from a very large central computer—and distributes heat to perimeter zones. Excess heat captured during the day is stored in large underground tanks that can be converted to chilled water tanks in the summer to reduce peak kilowatt billing rates. Heat from the computer center alone reduces the annual utility consumption by 20 percent. From the time the building opened in the winter of 1981, no fossil fuel heating has been needed.

Recycling rejected heat at PPG's corporate headquarters in Pittsburgh (1980): For yet another company with a vested interest in careful energy utilization, Mr. Mass devised a heat recovery system that gathers rejected heat from the computer center, recycles it through the refrigeration plant's condenser water, amplifies it with two industrial heat pumps, and then employs this water in the structure's perimeter heating system. PPG's all-glass-facade building, designed by Philip Johnson/John Burgee Architects, relies on Mr. Mass's heat balance analysis, which indicated that reclaimed internal heat would satisfy the building's heating requirements at outside temperatures of 3 degrees F. and up during hours of full occupancy. Supplementary heating would be required only when outside temperatures fall below that point during occupancy.

Using air in atriums (1965): The first atrium in a major public space was created for the celebrated Ford Foundation building in New York City, a revolutionary design by architects Kevin Roche John Dinkeloo & Associates. Mr. Mass developed a system that uses waste energy to maintain positive pressure so as to prevent infiltration from the outside.

OTHER EXAMPLES OF IMAGINATIVE, VENTURESOME WORK

Something has to be said for an engineer who is consistently asked to participate in buildings that push the frontiers of the building science. Marvin Mass has worked on controversial buildings such as the John Hancock Tower in Boston, and on some very modest buildings, such as one of the first solar houses, a structure developed by the University of Delaware in Newark to research photo voltaic and thermal collectors. At the other end of the spectrum, he is currently the engineer for two "world's tallest buildings" conceived by two daring, competing developers: Donald Trump and Harry Grant. Other of Mr. Mass's adventures in engineering include the following examples:

The Knights of Columbus headquarters in New Haven, a unique 80-foot-square building by architects Kevin Roche John Dinkeloo & Associates, appears to have no ducts or air diffusers. In fact, the structure itself carries air through the building via pairs of beams that create natural ducts along their lengths. Air is also transported and distributed through the floor deck. Cross beams on the exterior cut off the direct rays of the sun and function as louvers.

Mr. Mass was one of the first to get into the now-popular technique of "daylighting"—finding ways to utilize daylight from an atrium or through outside "light shelves" to transmit natural light deep into a building's interior spaces. One such building, the GSIS headquarters in Manila, the Philippines, designed by The Architect's Collaborative, reduced artificial lighting by 40 percent.

In Glasgow, Scotland, the St. Enoch Square Mall will, once completed, be the largest passive-solar, glass-enclosed building in the world. GMW Partnership of London designed the complex. A glass envelope protects the structure, which is naturally heated by the sun and naturally cooled through air ventilation.

"Cogeneration" first became well-known by the public in the last 20 years. One of the first and, at that time largest, total-energy plants in New York City is at the Kings Plaza Shopping Center, designed by Emery Roth & Sons. Its 12,500 KVA is generated on-site and re-sold to individual users.

Double-decker elevators are becoming essential as we build super-skyscrapers. For the Time-Life Building in Chicago, as early as 1965, Mr. Mass created and powered the first such elevators for the architect, Harry Weese & Associates, thus saving a considerable amount of shaft space – not to mention a lot of time for the unknowing passengers.

TEACHING

"I am not a professor not a researcher, but a practicing engineer," Mr. Mass says frequently. 'His modesty aside, for nearly a decade he has commuted from New York to Cambridge, Massachusetts to teach graduate design students at Harvard about "The Spatial Implications of Mechanical Systems." Since few design schools integrate engineering courses to this extent, the program is significant.

A 1948 graduate in Mechanical Engineering from New York University, Mr. Mass has regularly lectured at the University of Pennsylvania, Pratt Institute, and Cooper Union.

For four outstanding decades, Marvin Mass has set the pace for the profession and the industry. He is a member of the National Society of Professional Engineers, the New York State Association of the Professions, and the American Society of Heating, Refrigeration, and Air Conditioning Engineers.

But perhaps—for all his single-minded dedication and activity—his most aptly described by this citation from the New York Chapter of the American Institute of Architects, which in 1986 made him an Honorary Member:

> "With his remarkable sense of how buildings live and breathe, he patiently and generously makes us better architects."

In many ways, Marvin Mass's life reflects the quiet leadership that marked Franklin Pierce Brown's, coincidentally in an allied profession. The award in Mr. Brown's memory recognizes "meritorious improvements in the building and allied industries." Certainly, Marvin Mass can be considered to have made this his own life's work.

When I told my wife, Ruth, that I was to be honored by the Franklin Institute, she thought I was kidding. "There's no way you're getting the same award as Marie Curie!" But it was true. Ruth couldn't believe it; neither could I.

Excerpts from some of the letters submitted to the AIA in support of the Institute Honor conferred on Marvin Mass in 1990

EDWARD LARRABEE BARNES: Marvin Mass is a warm human being—a delight to be with. He has countless architectural friends who have a high regard for his warmth, humor, and friendship. He has a real understanding of design issues. He is also a damn good engineer.

JOHN BURGEE: Marvin can always be counted on to provide the best and most innovative solution to any complex mechanical engineering problem. His skill and talent in advancing the state of mechanical engineering are acclaimed by his peers in both engineering and architecture. Marvin is an essential part of the team that is necessary to successfully design buildings that best fit the needs of clients and users.

NORMAN C. FLETCHER: Marvin has been for me one of the high priests of the mechanical engineering profession. Whenever we have a significant commission demanding creativity as well as experience, we are apt to look for Marvin Mass and his associates.

ROBERT F. FOX, JR.: Generally, the buildings that are designed with Marvin's participation are unique. Because of his design sensitivity and his ability to listen to and understand the architect's language, he is able to produce extremely creative engineering solutions. I know of no other professional engineer who even approaches him in creativity, generosity, and sensitivity. He has shown us all a total commitment to his profession and to humanity.

JAMES INGO FREED: Mr. Mass is an unusual engineer in that while he performs his function with great excellence, he understands that his work must augment the "architecture" of the project in hand and not overwhelm it.

KEVIN ROCHE: Marvin is certainly one of the most imaginative and creative mechanical engineers working in the field today, with a long list of outstanding successes to his credit. His knowledge and interests go far beyond the working level of systems, and his creative mind continually searches for new and more effective ways to create and sustain the indoor environment.

PETER SAMTON: Marvin has made unique contributions as an engineer to the profession. Through his efforts, his firm has broken new ground in a host of areas and has immensely helped the architectural community face the current era of "smart buildings" and high-tech services. Marvin has been unselfish in communicating his ideas to the entire architectural community.

CÉSAR PELLI: Marvin is an exceptional mechanical engineer. I have worked with him on different projects for some 20 years now and I am always impressed by his knowledge, his clarity of mind, and his good humor.

HELMUT JAHN: The concept of many of our designs would not have been realized without the knowledge, direction, and enthusiasm brought to each project by Marvin. Many of the architectural details of our projects emerge from the integration of solid design and innovative engineering principles. The creative engineering abilities of Marvin Mass will lead us into the future and into new design directions that will emerge from new challenges.

THE AMERICAN INSTITUTE OF ARCHITECTS

IS PRIVILEGED TO CONFER THIS

1990 INSTITUTE HONOR

ON

MARVIN MASS

TEACHER, INVENTOR, ENGINEER,
A MAN DOUBLY BLESSED WITH WARMTH
AND INTELLIGENCE, A DESIGNER WHOSE
INNOVATIVE SOLUTIONS TO THE MOST COMPLEX
ENGINEERING PROBLEMS CAN BE FOUND IN MANY OF
THE NATION'S LARGEST AND SMARTEST BUILDINGS,
YOU HAVE BEEN WIDELY ACCLAIMED BY YOUR PEERS,
ARCHITECTS, ENGINEERS, AND BUILDERS ALIKE.
WHAT SETS YOU APART ARE THE OLD VERITIES:
RELIABILITY, CANDOR, A FIRM RESOLVE TO
NEVER STOP LEARNING, AND ONE QUALITY
YOU HAVE EMBEDDED IN THE MINDS OF
YOUR MOST PRECIOUS OFFSPRING,
YOUR STUDENTS: TO STRIVE
FOR WHAT YOU ONCE CALLED
'A BEAUTIFUL SOLUTION.'

MAY 1990

DAVID CHILDS, *Design Partner, Skidmore, Owings & Merrill*

Happy Birthday! Welcome to the club where all of the members think that anyone less than 50 years old is just a whippersnapper.

When I moved to New York to start my second career with Skidmore Owings & Merrill, I was met practically at the train by the best new business developer in the world: Marvin Mass. Not only did I do my first job in New York with you all, but so many others since. I'm pleased with the way that collegial relationship has continued with Cosentini, and happy that so many dear personal friendships have developed from it. So thank you, Marvin. Thank you, Cosentini, and many, many happy returns.

JOSEPH FLEISCHER, *Management Partner, Polshek Partnership Architects*

Marvin Mass is perhaps the most thoughtful and innovative engineer in the profession. His brilliance has changed the way we do architecture and his collaboration enhances the quality of the environments we seek jointly to create. Ultimately what matters most to us as architects is his commitment to quality and understanding and responsiveness to our design visions.

Among design professionals, Marvin leads the way in areas of charity and community involvement. All this, while raising a huge (by twentieth-century standards) family. Perhaps his wife, Ruth, is the one who really deserves the credit.

BOB FOX, *Founding Partner, Fox & Fowle Architects*

I met Marvin in the mid '70s. He was incredibly generous to offer us office space and then he helped us with clients and helped us to get work. He was so supportive, always there when we needed him.

Marvin was our personal engineer on all of our early projects, and worked closely with us in designing all the systems, the plumbing, the electric, everything to do with the building. He was hands on, just terrific. And we've enjoyed that association ever since. Cosentini has done a terrific job for us on many, many, many projects.

PHILIP JOHNSON, *Principal, Philip Johnson/Alan Ritchie Architects*

More than a generation ago I started to work for Cosentini; it's supposed to be the other way around! It seemed perfectly clear to me that we were meant to work together. And every time we did, we had fun and we always produced a good building. Now, why is that? It's got to do with a lot more than just mechanical engineering.

Marvin is an expert architectural advisor on the engineering side of building, and I always consult him whenever I can. These days projects aren't coming in quite so often as before. Why should they? I'm getting older now, so I don't get a chance to see Marvin as much. But I would see him every day if it were my choice.

On the occasion of Cosentini's 50th Anniversary in 2002, here's what a few of our friends had to say.

GENE KOHN, *Chairman, Kohn Pedersen Fox Associates*

KPF has had a fantastic relationship with Cosentini for most of our 25 years. We always enjoyed Marvin visiting, talking with us, giving lectures, and helping our people to understand the value of mechanical and electrical engineering. In fact, what Marvin did was convince most of our young architects to become engineers! Just kidding, Marvin.

When I think of New York engineers, Marvin is the first who leaps to mind. And now that Doug has joined the firm – I think he was born there – Cosentini will remain strong. We look forward to being on hand to celebrate your 100th birthday. Congratulations and best wishes.

RALPH MANCINI, *President, Mancini Duffy*

This is a tribute to Cosentini, and to Marvin and Douglas. I think the success of the firm reflects Marvin's attitude about life. And, being friends and business associates for many years, we've shared great and fond memories. Some of the finest projects we ever accomplished were executed as a team with Cosentini.

I hope the next 50 years will be as successful as the first. May you be healthy and prosper. It's been great to be associated with such a fine bunch of people. I thank you for the opportunity to express my feelings and my thoughts.

JAMES POLSHEK, *Founder, Polshek Partnership*

I would not miss the opportunity to congratulate Cosentini, and Marvin, you in particular, whom we've worked with for so many years. Shortly our firm will be 40, so we're just kids compared to Cosentini.

You have always brought a certain degree of levity, and therefore joy, to even the tensest of our many friendly architecture-engineering sessions. We're going to use you guys for another 50 years. Happy, happy birthday on this extraordinary occasion.

ROBERT A.M. STERN, *Dean, Yale University School of Architecture*

Congratulations! 50 years! That's a long time! We've worked together with Cosentini for a good 10 to 15 years, and it's been great to be associated with such a wonderful organization with such wonderful people.

Of course, I love Marvin Mass. To his enormous credit, he makes mechanical engineering, the dismal science, seem almost fun. And Douglas is learning. So let's keep going, threading those pipes through my beautiful spaces and not making a mess of them. Onward and upward!

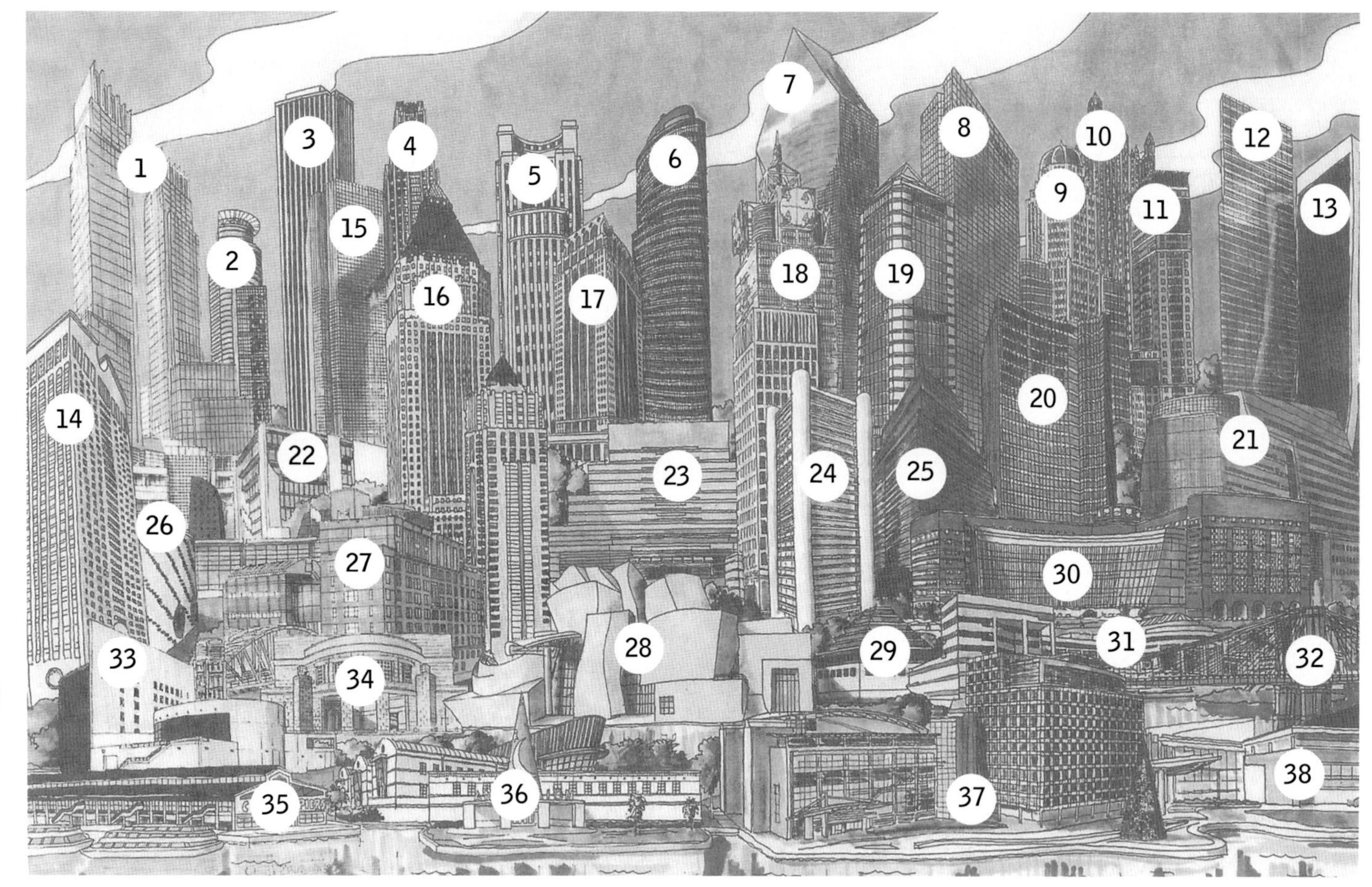

KEY TO SOME OF MARVIN MASS'S MOST PROMINENT BUILDINGS

1. Time Warner Center, New York, NY
2. First Bank Place, Minneapolis, MN
3. Amoco Building, Chicago, IL
4. Bell Atlantic Tower, Philadelphia, PA
5. Daniel Patrick Moynihan United States Courthouse, Foley Square, New York, NY
6. 53rd at Third Avenue, New York, NY
7. Allied Bank Tower, Dallas, TX
8. John Hancock Tower, Boston, MA
9. CitySpire, New York, NY
10. PPG Headquarters, Pittsburgh, PA
11. Carnegie Hall Tower, New York, NY
12. Metropolitan Tower, New York, NY
13. 9 W. 57th Street, New York, NY
14. AT&T Building, New York, NY
15. Park Avenue Plaza, New York, NY
16. Worldwide Plaza, New York, NY
17. 450 Lexington Avenue, New York, NY
18. Condé Nast Building, New York, NY
19. Park Avenue Tower, New York, NY
20. Republic National Bank, New York, NY
21. Newman Vertical Campus, Baruch College, CUNY, New York, NY
22. Ford Foundation, New York, NY
23. Crile Building, Cleveland Clinic, Cleveland, OH
24. Knights of Columbus, New Haven, CT
25. 380 Madison Avenue, New York, NY
26. Owens Corning World Headquarters, Toledo, OH
27. Stuyvesant High School, New York, NY
28. Guggenheim Pavilion, Bilbao, Spain
29. Museum of Jewish Heritage, New York, NY
30. John Joseph Moakley United States Courthouse, Boston, MA
31. General Foods Headquarters, Rye Brook, NY
32. Brooklyn Bridge Centennial Celebration, Brooklyn, NY
33. John F. Kennedy Library, Boston, MA
34. United States Holocaust Memorial Museum, Washington, DC
35. Chelsea Piers, New York, NY
36. Disney Feature Animation Building, Burbank, CA
37. UBS AG Headquarters, Stamford, CT
38. W.W. Grainger World Headquarters, Lake County, IL

Artist: Stephen Syznal

Massopolis

Some of Marvin Mass's most prominent buildings.
A watercolor collage celebrating the 50th anniversary of Cosentini Associates, 2002

DAVID CHILDS, *Chairman Emeritus, Skidmore, Owings & Merrill*
After I moved to New York City in the mid '80s, I began to work closely with Marvin Mass and soon realized through him the importance his profession could, and should, play in the design of buildings. Until that time, most of the mechanical engineers I had worked with saw themselves as mere problem solvers, and for many of the smaller buildings I worked on in Washington, that was sufficient. Marvin opened my eyes to the creative potential of mechanical systems in the design of skyscrapers and other large, complex projects. He always took my questions to the next higher level of thinking. Together, we worked as true design colleagues. I have now partnered with the very finest building systems professionals around the world, but I have yet to find one who could surpass Marvin's intellectual and design creativity, and his fundamental contribution in a collegial team of designers.

BRUCE FOWLE, *Founding Principal, FXFowle*
Marvin was the first mechanical engineer I ever worked with. It was 1961 and I was on the design team for the New York Hilton Hotel as a right-out-of-school scared-out-of-my-mind draftsman. Marvin couldn't have been nicer. He took me under his wing and taught me all about how buildings work. It was an invaluable experience and a very fortunate circumstance for a young architect.

Seventeen years later, when my partner and I were starting our firm Fox & Fowle, Marvin offered us space in his office at 2 Penn Plaza. It was located on the twenty-eighth floor between the elevator cores with no windows. The second means of egress was a hatchet on the wall, but the space was furnished with drafting tables and it was *gratis*! I might never have had the nerve to give up my secure job and start the firm if it hadn't been for Marvin's generosity. I will be forever indebted to him.

Marvin continues to be an inspiration, and not just for his remarkable engineering talents. His energy, his many interests beyond his profession, and his joy of life are something for all of us to aspire to.

JILL LERNER, *Principal, Kohn Pedersen Fox Associates*
I first met Marvin when I was with Ellerbe Becket, in the 1990s. He was famous and brilliant, and I knew it was an honor to work with him. He had collaborated with all the very best architects, and was well known as a great problem solver and strategist. Marvin always came to the table with new ideas and innovative approaches, always respecting and informing the designer's concept. Over the years, he became a wonderful mentor to me. In fact, Marvin assisted me almost 18 years ago in moving to KPF, where I am now a principal.

When I was first invited to Marvin's annual party, I felt privileged and duly humbled to be among such distinguished company—everyone admiring of Marvin's talent, and of Marvin and Ruth's generous hospitality. Marvin and Ruth have always talked about their wonderful children and grandchildren, and it has been my pleasure to meet the extended Mass family. I hope all the kin appreciate how brilliant and renowned Marvin is as an engineer, and how special he has been as a mentor to so many professionals, always interested in their career as well as in their family life.

More Tributes from Friends

ALAN RITCHIE, *Principal, Philip Johnson/Alan Ritchie Architects*
I got to know Marvin when I was with Philip Johnson, as our firm did a lot of work with Cosentini. Now, practically every other week, he and I go to lunch, share thoughts, and catch up on people we know. Having spent a good deal of time together over the past 30 years I can truly say that Marvin is one of my best friends. I look to him often, a bit like a father. I love him dearly.

After Mr. Johnson died, it's interesting that I would turn to an engineer for advice in the architectural field, but Marvin was the perfect person, very professional, very willing to help. He always gave good counsel and supported me. Marvin genuinely respects architecture and works with architects to preserve their designs. That's unusual. Most mechanical engineers are just not part of the conversation; all they want to do is make the building systems work. By contrast, Marvin is concerned with all aspects of the space. Buildings function so well because of him, but you don't see his work or even know it's there. If you want to hide the various systems or put them in unconventional places for architectural design integrity, Marvin provides the solution. That's very special, but the delight is really the man. Marvin is pure energy and creative enthusiasm. He loves people and they love him back.

Whenever possible, my wife and I enjoy spending time with Marvin and Ruth. How they complement each other! They're a fantastic couple and have such an incredible family. We recently attended Marvin's 85th birthday party with all the children, grandchildren, and a few close friends. Marvin greeted me by saying, "Alan, you're part of our family." It was a great honor.

NANCY RUDDY, *Principal, Cetra/Ruddy Architecture*
Marvin has the unique combination of wisdom and gentle strength that allows him to walk into a conference room filled with diverse egos—bank presidents, developers, construction managers, architects—and immediately win everyone's respect. With intuitive calm and grace, he consistently defines salient issues and then brings creative resolution to complex matters, making it all seem so easy!

I met Marvin when I was a young architect heading up a large, complex project. Wisdom might have told my employer that I was too young for the assignment. At Marvin's feet I learned everything I needed to know about New York real estate, managing powerful clients, and the integration of building systems in high-rise construction. Most important, Marvin had faith that I could do this well and gave me the courage to march forward. Now, 30 years later, with many projects under both of our belts, Marvin still cheers me on.

Everyone in the industry knows that Marvin is godfather to the architectural community. He has a passion for innovation and thrives on the intellectual debate and collaboration that make great buildings and great cities. With his sparkling blue eyes and friendly voice, he challenges us to be the best architect and the best human being we can be. Marvin makes people feel cherished and respected. His love of family, whether for his own kin or the many engineers and architects he has touched over the years, has created bonds within our industry that would not exist without his magic. Marvin and his wife, Ruth, bought my 3-month old daughter her first snowsuit. She is now 18, and there is never a time that I see Marvin, whether discussing sustainability or some other complex technological issue, that he doesn't ask after her with a twinkle in his eye. Marvin really knows what life is about.

Marvin A. Mass, The Invisible Architect

GETTING STARTED

Somebody once called me "the invisible architect," though, in fact, I am a mechanical engineer. Because I started so young and have lasted so long, I've had the chance to work with many of the great architects of the past half century and to help bring to life some of the world's most interesting buildings. My work does not necessarily "show," but without it, I humbly submit, these buildings could not function.

Most people haven't given any thought to the role of mechanical engineers and have no idea what we do. To put it briefly, it is our job to take an architectural scheme and give it life. The great fun I've had over the years has been to transform ideas into reality.

In the Middle Ages, the people responsible for new buildings were called Master Builders. They took on every aspect of a job, constructing a building from designs they themselves had created to meet the user's needs. Although the various functions that go into creating a building are now separated, I still try to do that: I find out what the user needs and then work with the architect and the other players to produce it.

Over the years, I have taught at Harvard, Yale, Cooper Union, the University of Pennsylvania, and other schools of architecture. In opening remarks to the first class of the semester, I'd describe the collaborative process like this: A building is more than skin and bones; it also has a heart, veins and nerves. Like a human being, it is composed of many parts.

Continuing the analogy, you, as architects, will determine what humans look like – the color of their hair and eyes, their height, weight, and facial features. The structural engineer will design the bones that support the whole, i.e. the columns, beams, slabs, and so forth. The rest of the body's functions—the heart, the lungs, the veins, the urinary track and even the brain—are the responsibility of the mechanical engineer. He designs the functions essential for the building's occupants, including heating, air conditioning, ventilation, plumbing, lighting, telephones, elevators, water supplies, sewage systems, security, fire protection, and facilities for the electronics that have become the brains of buildings.

A building may look beautiful, but to make it usable, the mechanical systems must work properly and function for their specific use. There are different requirements for a school than for an office, or for a museum, or a courthouse, or a concert hall, or an apartment building. The mechanical engineer, right from the very first day, must take into account the special needs of occupants in each type of building.

The mechanical engineer bears a lot of the responsibility for a building's success or failure. If the mechanics don't hold up or don't keep up with the times, the building is in trouble. If you hear that a building is being "upgraded," it's probably the mechanical systems being brought up to date, with changes that make older buildings more energy-efficient, and environmentally friendly. It is the heating, air conditioning, electric power, and computer facilities that are being improved, not usually the building's aesthetics.

SOME OF THE ARCHITECTS I'VE ENJOYED WORKING WITH:*

IN THE BEGINNING: Louis Kahn, Mies van der Rohe, Eggers & Higgins, Emery Roth, Eero Saarinen, Edward Durell Stone, Harry Weese

THE MIDDLE YEARS: Max Abramovitz, The Architects Collaborative, Edward Larrabee Barnes, Beyer Blinder Belle, Max Bond, Marcel Breuer, Gordon Bunshaft, Butler Rogers Baskett, Cambridge 7, Ivan Chermayeff, Peter Chermayeff, David Childs, Peter Claman, Henry Cobb, Alex Cooper, Aldo Cossutta, Lewis Davis, Davis and Brody, John Dinkeloo, Peter Eisenman, Howard Elkus, Elkus Manfredi, Ellerbe Beckett, Norman Fisk, Joseph Fleischer, Norman Fletcher, Norman Foster, Bruce Fowle, Bob Fox, Ulrich Franzen, James Ingo Freed, Joe Fujikawa, Gensler, Raymond Gomez, Michael Graves, Barney Gruzen, Jordan Gruzen, Charles Gwathmey, Wallace Harrison, Richard Hayden, Helmut Jahn, Gerald Johnson, Philip Johnson, Eli Jacques Kahn, Kahn & Jacobs, Kallman & McKinnell, Dan Kiley, Eugene Kohn, Costas Kondylis, John LoPinto, Raymond Loewy, Victor Lundy, Mancini Duffy, Mitchell Giurgola, Victor Gruen, Jordan Gruzen, Elliott Noyes, Gyo Obata, Derek Parker, Bill Pedersen, I. M. Pei, César Pelli, Perkins & Will, James Stewart Polshek, Lee Harris Pomeroy, Alan Ritchie, Crawford Robertson, Kevin Roche, Richard Roth, Moshe Safdie, Peter Samton, Michael Saphier, Gerald Schiff, Robert Siegel, Skidmore Owings & Merrill, Shreve, Lamb & Harmon, Frederick Stahl, Robert A.M. Stern, William Tabler, Edgar Tafel, Marilyn Jordan Taylor, Emanuel Turano, John Carl Warnecke, Welton Becket

MORE RECENTLY: Brennan Beer Gorman, Santiago Calatrava, Cetra/Ruddy, George Cruz, Steven Davis, Wendy Evans Joseph, Frank Gehry, Art Gensler, Peter Gorman, Zaha Hadid, Gary Handel, Hugh Hardy, Walter Hunt, Jill Lerner, Kallmann McKinnel & Wood, Kohn Pedersen Fox, Richard Meier, Didi Pei, Sandi Pei, Rafael Pelli, Renzo Piano, Richard Roth Jr., Swanke Hayden Connell, Rafael Viñoly

Interestingly, within this group are fathers and sons: Barney and Jordan Gruzen; César and Rafael Pelli; Lewis and Steven Davis; I. M., Didi, and Sandi Pei; and three generations of Roths: Emery, Richard, and Richard Jr. If I missed anyone, I'm sorry.

*See endpapers for firm names and other architects who have worked with Cosentini.

My career would have been difficult to predict in the 1930s in the working-class neighborhood in the Bronx where I grew up. I still see some of the people who were my friends back then – the ones who are still alive, that is. Though we came from modest families, barely surviving during the Great Depression, many of us did well in our lives. One boy went into law, and a couple became engineers or business owners. Our parents were first-generation immigrants, mostly from Eastern Europe, who worked hard, held us to high standards, and, above all, stressed learning: "Go to school. Get an education."

I went to New York's Stuyvesant High School, which is still doing a terrific job of educating and encouraging bright kids, many of them from present-day immigrant backgrounds. Then, when I was fifteen or sixteen, I went to the College of the City of New York, the justly famed City College. I was young because there was then a system in New York schools called "rapid advance." It was not that we were so brilliant but, I think, because of World War II, the city needed to get kids educated quickly.

High school took me two years, college three —very different from the now-standard four years for each. Before I finished City College in 1945 I went into the army. When I got home, the GI Bill helped me, like so many others of my generation, to go to a private university, what was then New York University's School of Engineering, where I got my degree in 1947.

I had a wonderful father who taught me to respect other people, to tell them what it was that they did well, and not to criticize. I've tried to do that all my life, though I don't always have his patience. I certainly have had an easier time than he did. My father faced a lot of problems because of the Depression. He would take whatever jobs he could find to support my mother, my sister, and me. But the worst of it was not the economics. It was that my mother died when I was seven. That was a terrible loss and it took several years for the three of us, together with my father's new wife, to share a home again.

When I graduated from NYU, the country was just recovering from World War II. I was not exactly well-connected and had no idea how to get started or where I would work. But I was too naïve to realize how much that mattered, so I turned for help to the newspaper's want ads. Miraculously, I found a job as a draftsman and engineer at Jaros, Baum & Bolles. The firm still exists, still does great engineering, and is still a model I try to emulate.

One day, in 1949, after I'd been working as a draftsman for a couple of years, Mr. Baum asked, "Young man, can you use a slide rule?" I answered, "I graduated from college. Of course I know how to use a slide rule."

That's how I got to work as Project Manager for Lever House on Park Avenue in New York, a building that was a pioneer of the International Style glass box. It's a classic that

Lever House, New York (Gordon Bunshaft/SOM, 1952)

looks great after its recent renovation and still works well today. A project manager coordinates the work of the architect, the structural engineer, the mechanical engineer, and other consultants. In effect, he puts together the designs of all those people. I guess Mr. Baum had been observing the work I was doing from the time I came on the job, so that, slide rule or no, he learned to trust me. My desk was only a few inches from his office, after all, but at the time it seemed to me a pretty swift decision.

Mr. Baum was a terrific boss who shared with me what he knew about the technicalities of building. He would explain how big a building could be, how many square feet, how many tons of air-conditioning. He had terrific instincts. When he looked at an architect's plans he could tell from experience rather than complex calculations what we would be dealing with.

Gordon Bunshaft of Skidmore, Owings & Merrill (SOM) was the architect for Lever House, the first major curtain wall building in New York. The design replaced, with blue-green glass and steel, the usual heavy masonry walls that skyscrapers had used before. And because the windows were fixed in place, it became the first mechanically climate-controlled and properly ventilated building in New York. In other words, since the windows could not be opened and shut as they could in the past, the interior climate was controlled by the air-conditioning and air-circulation systems, most of which were installed in the top three floors of the 24-story building.

I also was Assistant Project Manager on an insurance building then being constructed at Broadway and Fifty-ninth Street. The project manager on that job was William Randolph Cosentini, a wonderful man. I think the most important thing I learned from him was that you could be a nice guy in a tough business and still succeed. He was so generous we called him Benny after "Benny the Bum," a Depression-era comic strip character, who was always giving money away.

99 Park Avenue, New York (Emery Roth, 1954)

When Mr. Cosentini left the firm to start his own company, he asked me to join him. William Wuhrman and Edward J. Losi, engineers from other firms, came along as partners. Since I was then earning thirty-five dollars a week, I had little to lose. I had by then passed the Professional Engineer's exam, but the Board of Regents of the State of New York required that engineers be at least twenty-five years old to get a license. I turned twenty-five in 1952, got my license the moment I could, and became a partner in the firm, W. R. Cosentini & Associates.

The first job our new firm got was 99 Park Avenue. It was not unusual for someone considering starting his own firm to ask a client, "If we go out on our own, do you think you would have work for us?" The answer invariably was, "Of course," and the result almost invariably was that the work happened to have been assigned to someone else the day before your own company started up.

We had asked John Tishman, the well-known builder and developer, if we could do one of his jobs and he had replied, "Without a doubt." The day we opened up shop John called and said, "Here's your first job." Not only was it our first assignment, the 26-story aluminum-sheathed building was, like Lever House a couple of years earlier, one of the first curtain wall buildings erected in New York (1954). Imagine how amazing it seemed at the time: a skeletal steel-structure building whose skin was pre-assembled and stored right in the street at Park Avenue between 39th and 40th streets. The wall went up in 2-story units without any scaffolding, and the building, magically, was completely enclosed in just six and a half days. I still thank John for the favor he did us and he still eggs me on about this or that aspect of our work.

Then, very soon after we got started, disaster struck. Benny Cosentini died, at age forty-one, of a ruptured appendix. Ironically and tragically, his wife had died of the same cause five years earlier. Benny left two children with whom I kept in contact in the years that followed. Every once in a while, to this day, I get a letter from one of Benny's grandchildren saying: "Tell me about my grandfather."

By the time Benny died, I was one of three partners running the company. We decided to keep his name on the firm. We've been Cosentini Associates ever since, partly for tradition, partly to honor so generous a friend. He put me in business, kept me in business, and I don't really need my name on the door.

Benny Cosentini

Central Park West, New York City

THE EARLY DAYS

Although I have worked all over the country and the world, New York is my beloved home and it is where I got my professional start. The New York City that we know was built by architects over the last one hundred and fifty years. But from 1946 to 1980, the city was almost totally rebuilt to become the most important commercial and financial center in the world.

The architect most responsible for that triumph, the one who did most of the residential work, is someone most laymen have never heard of: Emery Roth. He died in 1948, and his sons, Julian and Richard, and eventually his grandson, Richard Jr., continued the work of the firm. Unfortunately Emery Roth & Sons went out of business in the late 1990s, but if you look at Park Avenue and Fifth Avenue and Central Park West, you may not see attention-getting architecture, but you will see buildings that "fit" New York, defining its streets and skyline with office buildings, hotels, and great residential blocks. Many of these buildings are similar but they make for a unified and dignified and eminently livable cityscape. Were the Roth buildings spectacular architecture? No, but they are as functional and usable today as they were when they were built fifty or sixty years ago.

Top: Zeckendorf's circular office;
Bottom: 383–385 Madison Avenue with Webb & Knapp's penthouse headquarters

In the 1950s Roth designed twin buildings at 579 and 589 Fifth Avenue, at Forty-sixth and Forty-seventh streets. The owner of the building in between would not sell, hence the separation. It was our second major job and our second with Emery Roth. We got the work because we were establishing a reputation for being able to devise flexible mechanical designs that could accommodate multiple uses. The twin buildings have now received new glass facades but, in fact, the design was never outstanding and still is not. The space between the floors is only about eleven feet, which leaves very little room for mechanical systems. Today, we typically have 13–14 feet between floors.

In addition to the work of Emery Roth, it was the visionary developer William Zeckendorf, of Webb & Knapp, Inc., who was responsible for revitalizing New York with new office buildings. Much of the development of Sixth Avenue and lower Park Avenue came through his initiative. And, from an architectural standpoint, his insistence that more natural light be brought into the offices reinforced Modernist ideals and helped promote the glass-walled buildings of the period.

Webb & Knapp's headquarters on the top floor of 383 Madison Avenue was redesigned by Zeckendorf's in-house architect I. M. Pei in 1949–52. Since everything revolved around Zeckendorf, Pei reasoned, and since "it would be ridiculous to create any environment for Bill other than one consisting exclusively of himself," Pei reinterpreted the traditional corner office as a free-standing teak cylinder,

25 feet in diameter, like a headquarters within a headquarters. Light entered the wooden drum through a continuous transom and from a shallow dome equipped with colored "mood lights" above Zeckendorf's desk. When Bill was angry at someone, the lights would shine red. If he liked you, he'd change the lights to blue or pink.

It was Zeckendorf who acted as the broker in putting together the 17-acre site for the United Nations Headquarters, which was constructed in 1949–1952 in an old slaughterhouse district along the East River in Manhattan. Zeckendorf sold the property to David Rockefeller for "any price he wanted to pay," which turned out to be $8.5 million. The story—apocryphal or not—was that Bill added the proviso that if the UN ever relocated, he would retain the privilege of brokering the property.

Zeckendorf, big and heavy, almost always had a large hat on his head, a cigar in his mouth, and a phone to his ear. He was a warm, kind man and very adventurous. Once, flying over Montreal, he pointed out the window and said to Harry Cobb, "See that big hole? [an open cut exposing the tracks of the Canadian National Railway]. That is where I'm going to build." And it is on that site that Cobb designed, and Zeckendorf produced, the well-known Place Ville Marie project that sparked the Renaissance of downtown Montreal. We worked on this 42-story project, which is still very much in use today.

It was the first major building in Cobb's long career. Instead of the small residential commissions of most beginning architects, Harry launched his practice at age 29 with a 3.7-million-square-foot landmark.

In the 1950s, during the early days of my firm, jobs at the big, fancy headquarters like AT&T, U.S. Steel, and General Motors were going to my better-established competitors who were working for well-known architects. My work was coming from new developers who, in those days, were called speculative builders—the Fishers, Kaufman, LeFrak, the Resnicks, Tishman, Uris, and, of course, Zeckendorf.

We would work on a building that Emery Roth designed but for which there were no tenants lined up to rent space. The developers undertook projects without a particular user in mind, but with designs flexible enough to accommodate the different needs of various tenants. Corporations soon discovered these flexible

William Zeckendorf with an early car phone

new buildings and realized that they could be adapted with great economy and efficiency for their own specific needs. Little by little, our firm began to work on corporate headquarters located within the spec office buildings. After that happened, fewer and fewer single-occupant buildings were erected. It was too expensive a way to operate and the building would end up being too difficult to convert to another use.

For example, for the headquarters of CBS on Sixth Avenue and Fifty-third Street in New York (see Chapter 4), the system we designed supplied air from machinery on the second and the thirtieth floors. That's fine for one corporate tenant, but the system does not work if there are multiple tenants with multiple requirements in the same building. Let's say a tenant on the fifth floor deals with world financial markets and needs to start the day at 3:00 a.m. You would have had to turn on the air-conditioning system in the entire building to accommodate him. All the floors would have to work at the same time.

Because spec buildings were designed for multiple tenants or for people within the same firm but with differing hours or other needs, the building type was found much more versatile and has stood the test of time.

One way we were able to make flexible design possible was to work with the manufacturers to produce modular rather than central building systems. This method made the building more energy efficient and eliminated the need for complicated, cumbersome, and very expensive equipment. It also meant your office could be warm in winter and cool in summer without requiring building-wide synchronization. And because use was individually metered, you could monitor what you were spending and thereby keep an eye on costs. It sounds so obvious now, but when we started, this was a revolutionary idea.

It was in the 1950s and '60s that the emergence of these new systems became important. But on other fronts, it took me a long time to convince the builders and architects that all parties—including the structural and mechanical engineers, architects, and contractors—should be in on the planning from the very beginning of a project.

Certain buildings constructed before 1970 did not take into consideration the space required for mechanical systems. One architect came up with complete plans for a church, and when I pointed out that he'd left no room for shafts and pipes and ducts, he said, "That's all right. We'll put them in the corridors." If, for another example, you go to Cooper Union at Astor Place in New York, you'll see radiators standing in front of the windows, instead of below them, because the architect didn't leave enough room and proceeded without speaking to the mechanical engineer (not me).

Once, when I was teaching at Yale, I took a group of architecture students to the Knights of Columbus headquarters in New Haven, on which we had worked (see Chapter 4). We went into the emergency generator room in one of the corner towers. The students were amazed that there was only about six inches of clearance between the wall and the equipment, and they jumped with fear when the generator started up. "Now you know," I said. "When an engineer asks for more space, give it to him."

In steel buildings, the structure goes up and we install our ducts, wires, and pipes beneath it. But concrete construction is different because it's a monolithic mass. It was Carl Morse, who came up with the idea of burying the electric conduit within the concrete slab itself rather than having the wiring on the outside as had previously been done. Carl was a pioneer in construction management, and was widely respected for his work on the Pan Am Building, the Helmsley Palace Hotel, and other Manhattan skyscrapers. When he died in 1989, the *New York Times* saluted him for having "helped shape New York's skyline."

Carl had started out working exclusively on residential buildings, and then one day he was hired to do an office building. He came to me and asked, "Marvin, what's the difference between the two?" After twenty minutes of my illuminating lecture, Carl became an expert on the subject of office construction and went on to become the head of Morse Diesel, one of the largest building construction companies in the world. As a matter of fact, Carl himself was a one-man construction company. All the trades working on a site would report to him individually, and he left little room for slack.

An electrical contractor told me that he once overheard Carl tell a union man on the job, "If you don't get your conduit in today, I'll pour the concrete over your feet." Carl's reputation was such that John D. Rockefeller III asked him for help on Lincoln Center, saying he wanted "a real S.O.B." to take charge and get the multi-building project built. Carl said he'd do it for free if Rockefeller called him "Mr. S.O.B." And so he did. And Carl spent the next 20 years as an unpaid consultant at Lincoln Center.

There was another important change that Carl Morse came up with. In the old days, wooden forms were set between the beams of a building and the concrete was poured into the forms. You had to wait until the concrete hardened, then remove the forms, and hope the concrete was smooth. Carl decided to use "metal deck" or fluted forms, then pour the concrete, and leave the steel in place. The conduit for wiring then goes between the forms. Most buildings in New York now use this method.

Carl Morse's distinguished career spanned the birth and maturation of modern American architecture. Throughout his six decades of work, architects and MEPs recommended new approaches, new building systems, and equipment, and Carl invariably called to ask my opinion, even when we at Cosentini were not the engineers. I was always happy to help.

Left to right: Harry Weese and Mies van der Rohe, Louis Kahn, and Ed Stone

WEESE, MIES, KAHN, AND STONE

In the early 1950s we were a young firm looking for work. A builder might recommend us or maybe an architect, and we would jump at the chance for a job. Looking back, I see just how remarkable it was that well-established architects willingly worked with a young, relatively untested engineer, but at the time I did not stop to appreciate my situation. I was new in business and clients were coming along: Weese, Mies, Kahn, Stone, Eggers, and Emery Roth. Great! Now, of course, I look back at those days with amazement.

Although I never worked directly with Frank Lloyd Wright, I worked during my early years with Edgar Tafel, Wright's leading disciple, on a new Church House for the First Presbyterian Church on Fifth Avenue in Greenwich Village. The church was established in New York in 1716 and has occupied its present site since 1846. Though the building Tafel designed is modern, it harmonizes with the eighteenth-century brownstone church because its exterior is almost exactly the same color. From my perspective, Tafel stood out because he was one of the first architects to actually think about the mechanical systems that were being developed and to incorporate them into his plans.

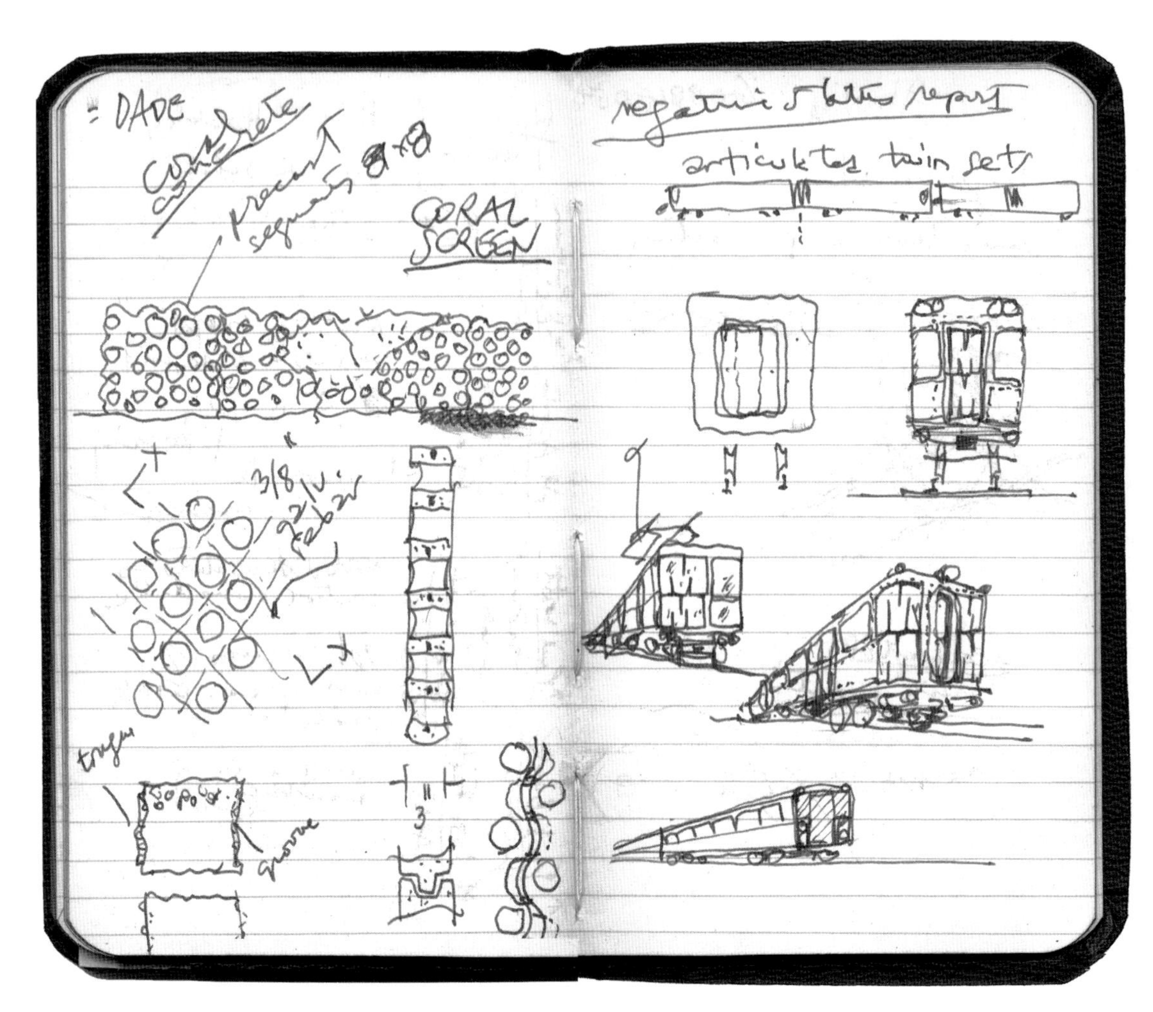

Harry Weese was involved in every aspect of the Washington Metro project, as his sketches for precast concrete segments and rail car designs suggest.

HARRY WEESE

Harry Weese was a Chicago architect who in the 1950s and '60s was becoming a major figure both in new buildings and in the restoration of historic architecture. Over the course of a half century he designed nearly a thousand buildings and played a major role in the post-war development of Chicago. Harry was very important to me throughout my career. It was through Harry that I got to work on the Chicago Field Museum restoration, the Metro system in Washington, D.C., and many more projects, including a few in Columbus, Indiana.

Weese was born in 1915 and died in 1998 after a distinguished career that began at MIT, where he studied under Alvar Aalto. Weese then went to the Cranbrook Academy of Art in Bloomfield, Michigan, and studied city planning with Eliel Saarinen, Eero's father. At the time there was a pretty amazing group of students at Cranbrook, including, among others, Charles and Ray Eames, Florence Knoll, and Benjamin Baldwin, Weese's future business partner and brother-in-law.

Initially, we did only big projects for Harry in Chicago, but he felt guilty because local engineers were left with just the small jobs. So Harry said to me, "I'm not going to give you only the big jobs. There are a lot of smaller projects in Chicago that you also have to take on."

In the early 1950s we opened our first branch office in Chicago to accommodate Weese—and of course, to cultivate work with other architects. In those days, I'd travel to Chicago every Friday, meet with Harry in the morning, and then he'd take me out to lunch, a

very wet lunch. We'd sit in the restaurant, and while he drank his martinis, he'd make drawings on paper cocktail napkins to illustrate his ideas. Harry would sketch his own vision of Michigan Avenue, of an island that could extend into the lake, and other grand ideas, many of which came to fruition. I kept those napkins, and shortly after Harry died, I sent them to his wife, Kitty, for a book that was being written about Weese's life and work.

We worked with Weese on the Time-Life Building in Chicago, a 30-story tower of Cor-Ten steel and glass. It was among the first buildings in the country to use double-deck elevators. We knew this scheme would save precious space in the building but had no idea how it would work in practice. It turned out that at five o'clock every afternoon, all of the building's occupants left at the same time. The elevators did a great job of handling the high volume, but as all the people piled into the lobby, there weren't enough exit doors to discharge them. More doors had to be added.

Harry was an avid yachtsman and kept a boat on the lake. One day he invited Ruth and me to join him on an excursion. After we were out a couple of miles, the wind died down completely and Harry turned to me and said, "Okay, Marvin, you take over." I had never sailed a boat before. Finally, I got us going—under motor.

One of our more interesting jobs came in the late 1960s and the early 1970s, when we

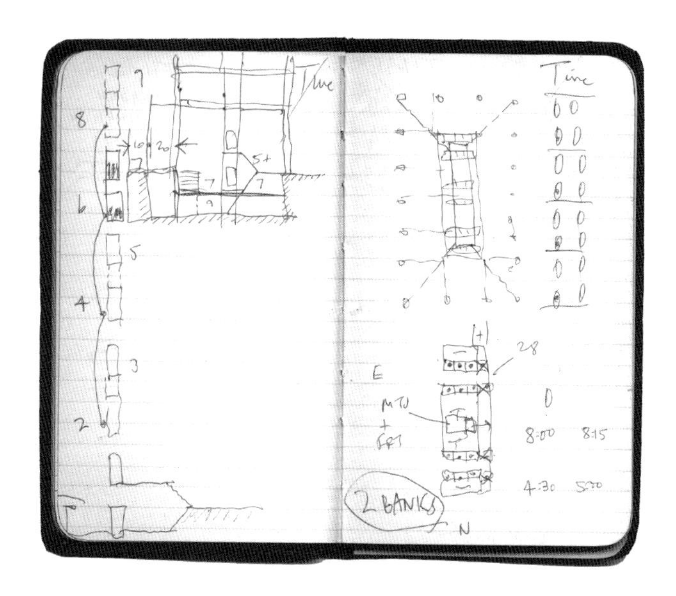

consulted with Weese on his masterpiece, the precast-concrete Metro system in Washington, D.C., which Harry designed in a marathon charrette over a long weekend. Harry thought that, to keep the stations climate-controlled, he would have to use double concrete walls with the ductwork running through. But instead, we suggested relatively simple, ten-foot hooded "air fountains" that stand on the station platforms every forty or fifty feet. The towers blow temperature-controlled air onto the passengers and then take it back through an exhaust system under the platforms. The towers are also used to post subway maps and directions.

Left: Time-Life Building, Chicago (1969)

Above: Weese sketches for the Time-Life Building's innovative double-deck elevators.

Right: Coffered Metro station with air fountain, Washington, D.C. (1969-1976)

The Chicago World's Fair of 1893 (before my time!), properly known as the World's Columbian Exposition, was dismantled in 1894, but one building was left standing. That grand structure was converted to Chicago's first natural history museum (now the Museum of Science & Industry). Because the city wanted a collection to rival those in the Smithsonian Institution in Washington, D.C., and the American Museum of Natural History in New York, their holdings rapidly expanded and they soon outgrew their quarters. In 1915–1921, Daniel Burnham's office erected an imposing new building named for the museum's first major benefactor, the department store magnate Marshall Field. All this took place, of course, before the advent of building-wide air-conditioning systems.

A half century later Harry Weese was engaged to renovate the building. Obviously, no space had been set aside for air-conditioning systems in the original structure, and therefore, we had to use the 50- to 100-year-old ventilation duct shafts. Strangely, it all worked out, and even today, the building still functions well.

The Field Museum, Chicago. (Office of Daniel H. Burnham, 1921; renovated by Harry Weese, 1977). The 300-foot-long, 70-foot-wide Great Hall houses "Sue," the largest and best Tyrannosaurus Rex fossil yet discovered.

LUDWIG MIES VAN DER ROHE

One of the bonuses of opening up shop in Chicago was that we got to work with Joseph Fujikawa and Gerald Johnson, members of the firm of Mies van der Rohe. Mies was one of the most influential architects of the twentieth century, like Le Corbusier before him. The famous phrase "Less is more" is his. I worked on a number of his projects but can't really say that I knew him. Every time I'd come to the office, Jerry would say, "Hey, Marvin, Mies just left." I could tell that was true because his cigar was still smoking on the table.

Although Mies was in the forefront of Modernism, his buildings were very traditional in their design. They were steel structures with hung ceilings and contained space for our ductwork, the need for which he understood. Unlike many architects of the time, Mies was not so arrogant as to say, "I want a twenty-foot ceiling but you can only have a few inches of that."

In 1962, in Baltimore, we worked with Mies's firm on One Charles Center, a 23-story building of dark brown aluminum, with a dramatic 2-story lobby. A predecessor of the iconic Seagram Building in New York, this landmark tower poured new life into Baltimore's urban renewal and gave the city its first great work of International Style architecture.

A similar building on which we collaborated a few years later was One Illinois Center, a 30-story black-glass-and-steel tower, overlooking the Chicago River. It was the first component of an 83-acre mixed-use development master-planned by Mies on the site of Chicago's old railyards. The building, which epitomized the International Style, was so successful that its design was duplicated for the nearby Two Illinois Center. Unfortunately, Mies died in 1969, while construction was under way, so the twin was completed by his firm.

Mies's buildings were sometimes criticized as being too similar, but they worked well and were commercially successful. The 38-story Seagram Building in New York, which I did not work on, was Mies's unchallenged masterpiece. It became the model for countless office buildings around the country, the absolute last word in corporate modernism. Among the many features that set this building apart was its bronze and tinted-glass facade (the first bronze skyscraper ever). Bold and beautifully detailed, the tower was unique on Park Avenue, where, aside from Lever House across the street, only masonry buildings had previously been constructed.

Seagram and Lever House also stood out as the only two buildings that did not occupy their full sites. An open air well was cut through the platform of Lever House to create a privately owned garden court for public use. Meanwhile, Seagram was set back ninety feet to create a huge plaza, so that about half of the space is left open for the public to enjoy. Reflecting pools were set into the plaza, and its low green marble walls (borrowed from Mies's Barcelona Pavilion) have become one of the most popular outdoor "sit-and-watch" places in New York.

Within the tower, Philip Johnson designed the Four Seasons, which quickly became one of the city's most chic dining places. Its Grill Room and main dining room are sheer elegance. In collaboration with Phyllis Lambert, daughter of Seagram's CEO Samuel Bronfman and an architect in her own right, Johnson transformed the restaurant into a veritable museum, installing a large Picasso in the corridor connecting the two sections of the restaurant and other major artworks in the lobby and throughout.

Years later, when my son graduated from Harvard Law School, I wanted to take him to lunch at the Four Seasons but couldn't get a reservation. I called Philip to see if he could help. About an hour later his secretary called to say I had a 12:30 p.m. reservation. As I walked in with my son, they ushered us to Philip's table, which was "reserved' for him every day unless he called to say he wasn't coming. Throughout lunch, people came over to ask after Philip. He actually came in later and sat somewhere else. [P.S. Philip paid for our lunch].

One of the great accomplishments of Mies van der Rohe was educating young architects. After he left Nazi Germany, Mies came to Chicago, hired by the Armour Institute of Technology in 1938. In 1940 Armour merged with Chicago's Lewis Institute, and IIT, the Illinois Institute of Technology, was born. Mies designed the original campus of the school and many of its buildings, though by the late 1950s, Skidmore Owings & Merrill had taken over.

Under Mies's direction IIT turned out some outstanding architects, including James Ingo Freed, Hans Hollein, Helmut Jahn, and Phyllis Lambert. Lambert was heavily involved with the school, and it was through her good offices that Mies, together with Philip Johnson, got the commission to do the Seagram Building.

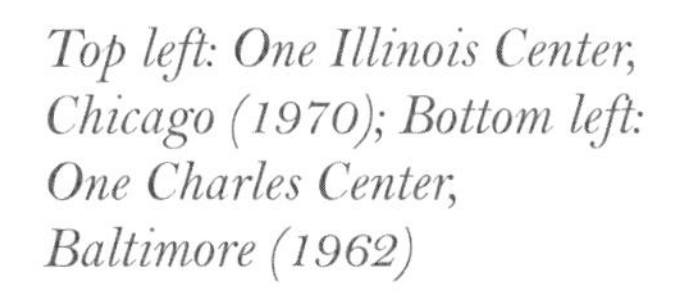

Top left: One Illinois Center, Chicago (1970); Bottom left: One Charles Center, Baltimore (1962)

Right: Mies van der Rohe and students

LOUIS I. KAHN

When Louis Kahn died in 1974, he owed my firm—among many others—a great deal of money. It was not until the release of *My Architect*, Nathaniel Kahn's wonderful film about his father, that I understood why. His complicated life, with several homes and families, required more money than most architects could make. I used to tell my students that if they wanted to go into architecture, they'd better find themselves a rich spouse.

Kahn hardly looked the part of Don Juan. He was short and pudgy, and had a badly scarred face (the result of a childhood accident). Born in Estonia in 1901, he came to this country with his impoverished family in 1906 and, on scholarship, studied architecture at the University of Pennsylvania, from which he graduated in 1924. In those days, it was difficult for an unprepossessing Jewish immigrant to find work. Kahn didn't receive his first major commission until he was 50 years old. Meanwhile he taught at Yale (1947–1955) and subsequently at the University of Pennsylvania, where he remained on the faculty until his death.

In the early 1960s, I was asked to give a series of lectures to the students at Penn about the coordination of the work of architects and engineers. Kahn would pop in from time to time and listen and preach to the class. (Keep in mind that there were almost no women in architecture at the time.) He would say, "Gentlemen, function is form. When you find out what the function of a building is, then you can design it. And when your engineer tells you he needs something, listen to him and design it into the building."

Kahn did not do many buildings in his lifetime. On that subject, I. M. Pei has said, "It is more important to design the three or four masterpieces that Louis did, than fifty or sixty buildings." Among his most notable are Salk Institute for Biological Studies in La Jolla, California; Kimbell Art Museum in Fort Worth, Texas; Yale Center for British Art in New Haven, and the remarkable Capitol Complex in Dhaka, Bangladesh.

Kahn had the reputation of being impatient and difficult with clients but I liked him a lot. He gave me, a young newcomer, a wonderful job in 1957 on the Richards Medical Research Laboratories at the University of Pennsylvania.

The building had four quadrants surrounding a central core. Each of the quadrants was about eighty by eighty feet. Kahn wanted this space to be totally unencumbered by columns or pipes so that, as the labs' needs changed, the partitions could easily be moved. He asked how I could provide all the mechanical systems and ventilation for multiple labs without interfering with the movable partitions. I answered, "Put the pipes on the outside of the building. That way, with the mechanical systems feeding in from the top, rather than in permanent spaces within the building, you can change the walls as you like." It was the first building I know of in which the mechanical systems were positioned outside the building.

Another challenge arose from Kahn's concept of servant and served spaces, without columns to support the eight-foot-high ceilings. To solve the problem, at about every six or eight feet, he used a concrete Vierendeel truss to span the entire eighty-foot-wide structure. The truss contained symmetrical openings through which our work could be installed.

One lab required an electric oven that was not near any wiring so I ran the cables on a metal tray through the truss. When Kahn saw it, he asked, "What's that for?" When I explained, he said, "Put another one on the other side, for balance." We didn't actually need another tray, but the architect wanted symmetry.

We also had a ventilation problem. Because Kahn had not given us enough space on the floor itself, we had to put a fan in a stairwell. This bothered me a lot but seemed to cause no concern for Kahn. He said again what I had heard so many times before, "Young man, function dictates form. You tell me what the building will be used for and I'll design around it." With a fan in the stairwell, I guess.

Entry porch, Richards Medical Research Laboratories, Philadelphia (1960)

Mies (left) and Kahn discussing form and function

At the same time, I was working in Chicago with the firm of Mies van der Rohe, who was famous for saying just the opposite: "Form follows function." For Mies, buildings were absolutes, pure geometric boxes that could be used for any variety of building types. It seemed to me that the problem with Kahn's concept was that if the function changes, the building often cannot be made to adapt.

Architects may design buildings that look great and even work wonderfully for the first user. But when a second user comes along, with different needs and goals, the original building construct no longer makes sense. It's one of the reasons Emery Roth's buildings in New York were so successful. They could easily be adapted to changing needs.

Kahn died of a heart attack, quickly and anonymously, in the men's room of Pennsylvania Station in 1974 as he was returning to Philadelphia from his government complex in Dhaka, Bangladesh. One of the greatest architects of our time remained in the morgue for three days before being identified. It was a sad end to his brilliant, difficult life.

Alfred Newton Richards Research Center at the University of Pennsylvania (Louis I. Kahn, 1960)

EDWARD DURELL STONE

Ed Stone was born in 1902, in Fayetteville, Arkansas, and went to the University of Arkansas. From then on his career was unconventional. He attended Harvard, left for MIT, graduated from neither but was awarded the prestigious Rotch Traveling Scholarship that helped broaden his architectural vision.

Stone's many accomplishments include his work in New York on Rockefeller Center, Radio City Music Hall, the original 1939 Museum of Modern Art (with Philip Goodwin), the General Motors Building, the Huntington Hartford Museum, and his own residence with its idiosyncratic grillwork on East Sixty-fourth Street. He also did the John F. Kennedy Center for the Performing Arts in Washington, D.C., the United States Embassy in New Delhi, and his incredible campus for the State University of New York at Albany.

Many of Stone's buildings were highly admired, and just as many were controversial. But Stone was not a man to listen to other opinions. He was a dictator. On every job I remember him saying, "That's what I want. That is what I want." Whether he was right or not.

Most of the architects with whom I have worked are committed to their own designs but they understand that the client's needs are of primary importance. Stone begged to differ.

For all that, Edward Durell Stone did some very good work and really wasn't responsible for some of the things he was blamed for. The fact is that some buildings don't work for reasons having nothing to do with the architect or the engineer. Stone's Huntington Hartford Museum at Columbus Circle in New York is a good example.

A few years ago I went to a public hearing at the Landmarks Preservation Commission, which was debating whether the building should be designated a landmark or whether unprotected, it could possibly be torn down. The new developers claimed that the museum wasn't a functional place to begin with: the galleries were small, circulation was difficult, and so forth. I was the original mechanical engineer on the project and can honestly say that none of the problems resulted from an error on the architect's part. The problems arose because the building wasn't made for you or me or for the modern museum-going public. It was built to house the private art collection of Huntington Hartford, the gadabout entrepreneur, impresario, man-of-leisure heir to the A&P supermarket fortune.

According to Frank Lloyd Wright, Mr. Hartford was "the sort of man who would come up with an idea, pinch it in the fanny and run." For his collection of nineteenth- and twentieth-century art he wanted a clear alternative to the Museum of Modern Art (which Edward Durell Stone had also designed, in 1937). The idiosyncratic museum in Columbus Circle challenged Modernist orthodoxy. It was what Huntington Hartford wanted and what he got.

When the Museum of Arts and Design took over the building as a general public gallery in 2003, it required big spaces and high ceilings. The function now changed completely.

I was disappointed with the radical exterior changes the building underwent, concealing its

The Huntington Hartford Museum, Columbus Circle, New York (Edward Durell Stone, 1964). Entrance arcade and exhibition gallery.

famous curved white marble facade and "lollipop" columns. The new exterior is a boring disaster, though I have to admit that the reconfigured interior space works very well. Stone's exterior was better-scaled and somehow more human; it had always just seemed an essential part of Columbus Circle.

Stone did not build anything else on Columbus Circle, although he had a grand vision to eliminate traffic and ring the circle with Doric columns salvaged from the Seventh Avenue facade of the old Penn Station. Things would have been very different if he had executed that concept.

In 1970 the former Gulf and Western Building, now the Trump Hotel, was designed by Tom Stanley of Dallas, Texas, who had no idea about New York or what the city wanted or needed. And to make matters worse, after the 45-story tower went up, it moved, a lot, in the wind—and that northwest corner of Central Park is a very windy corner. The structural engineer James Ruderman was brought in to redesign the core and make the building more stable.

My firm also worked on the new Time Warner Center that borders the west side of Columbus Circle. The original plan for the multi-story base of the complex was confused because there was no one person or firm commissioning the design. Each participant had his own ideas and it was a difficult problem to make all the requirements work together. There was also intense community objection. Jacqueline

Kennedy Onassis, for example, felt that the building's two towers would cast a long shadow on Central Park every afternoon at 4:00 p.m. Studies were done and the building had to be moved so it wouldn't block the sun. In all, it took nearly fifteen years, and several architects, before construction could begin.

Our offices had been in the building that was torn down for Time Warner, the old New York Coliseum Convention Center and Tower at 10 Columbus Circle. I know it was not an architectural masterpiece, but I liked the building, and you couldn't beat its location. Frank Lloyd Wright even showed it a little forgiveness. "It's all right for New York," he said, "but I hope it stays there."

Under David Childs, the Time Warner Center, like Stone's museum, made an appropriate response to the pivotal site, to the park, to the street pattern, and, most important, to the beautiful fountain in the middle of the Circle. I am interested in architecture that will still give pleasure fifty or sixty years down the road and that tries to do something new and innovative.

Those same criteria certainly apply to the State University of New York at Albany, which I worked on with Ed Stone in the 1950s and 1960s. For the new Uptown Campus, Stone grouped four residential quads around a central academic podium, consisting of thirteen 3-story buildings under a single overhanging roof supported by beautiful continuous arcades. The whole design had a very sculptural effect and, unlike traditional colleges with their scattered buildings in different styles, this campus was completely integrated and unified.

Left: Columbus Circle with the reclad Huntington Hartford Museum; Bottom: Time Warner Center (David Childs/SOM, 2003)

About twenty years after the buildings were erected, the university wanted to expand. Of course when other colleges have growing pains, they just erect a new building, but SUNY's campus was not open to change. The university asked me if I would speak to Mr. Stone and find out how they could incorporate a new structure into his enclosure. When I relayed the message to him, Stone answered, "Listen, if they want to build another building, let them do it next door and not mess with mine." SUNY's new Albany Nanotech complex was begun in the late 1990s, and is still expanding, to the west of Stone's intact campus.

Below: State University of New York at Albany, now the Uptown Campus (Edward Durell Stone, 1971)

Left to right: Kevin Roche, César Pelli, and John Dinkeloo (holdng CBS model); Kevin Roche and Eero Saarinen inspecting stone samples for the CBS Building; Eero Saarinen

SAARINEN, ROCHE, DINKELOO

If a critic were to ask me to name the architect I admire the most, I would answer Eero Saarinen without hesitation. Born in Finland in 1910, he was one of the greatest architects who ever lived. Eero came by it genetically, since his father, Eliel, was also a distinguished architect and the head of their joint practice. Among other projects, the two architects worked together in planning the 25-building General Motors Technical Center in Warren, Michigan. The 320-acre office campus, which Eero ultimately cast into an major icon of modern architecture, was a visionary model for corporate research and production facilities. It deserved its reputation as the "Versailles of Industry."

In the 1960s, a decade after Eliel's death, I went to Finland and visited Villa Hvitträsk, the family house designed by Saarinen senior, where Eero lived until he was 13 years old. I was struck by the fact that the furniture in this house was amazingly similar to that designed by Frank Lloyd Wright. It was almost unbelievable. The two men had lived thousands of miles apart and were very different in temperament and style. (Wright insisted that the only thing he "ever learned from Eliel Saarinen was how to make out an expense report." About Wright, Eliel gamely allowed he was really "a sweet man underneath.") For all of their differences, the furniture they designed and used had the same straight backs, the same clean lines, and the same respect for craftsmanship. Both had very much the same sensibility and the will to create a total work of art so that every object was an integral part of a larger context.

With a similar but more sculptural view of the whole, Eero Saarinen and his firm designed some of the great buildings of the twentieth century. Among the most notable are Kresge Auditorium at MIT and Ingalls Skating Rink at Yale, both innovative thin-shell structures of reinforced concrete. There was also Dulles International Airport, which Eero considered "the best thing [he had] ever done"; and the TWA Flight Center at Kennedy Airport, probably the most famous airline terminal in the world. Saarinen's success at General Motors led to other major corporate commissions, including the headquarters of CBS in New York and new facilities for IBM. I had the pleasure of working on all of the latter projects.

My association with Eero was very important to me. Though I was then the new kid on the block—young, with a new license and a new firm—John Dinkeloo, Saarinen's partner, heard about me from a contractor in New York. It was at a point when Eero was just making an independent name for himself, and I have to admit that it took me a while to understand that this wasn't just another architect that I was launching myself into partnership with. We were already working with Mies van der Rohe in Chicago, Louis Kahn at the University of Pennsylvania, and Harry Weese in Chicago. Saarinen led the next generation after those great architects and, of course, when that became clear, it was thrilling to work on his buildings, which were always innovative and challenging.

Though I had not yet met Saarinen, one day in the early 1950s, I was invited to a meeting at his office. It was at the back of an automobile dealership in Bloomfield Hills, Michigan, and when I got there five or six people—Kevin Roche and John Dinkeloo, as well as James Owens and Philip Kinsella, and a few others—were sitting around the table discussing one of Irwin Miller's projects in Columbus, Indiana. I kept my eye on the door throughout the meeting, waiting for the miracle man to join us. As it turned out, he'd been there all the time, sitting at the table, sleeves rolled up, just another participant.

The list of soon-to-be famous architects who worked for Saarinen is remarkable: Kevin Roche, John Dinkeloo, César Pelli, Robert Venturi, and many others who got their early training as designers, or preparing models or presentations. Kevin Roche said of these years: "Everything I know, I learned from Eero really. I learned his approach to things: how to work, the consistency, and the endless 'never give up' sort of attitude; you keep at it, keep at it, keep at it, until you feel that you've got it right." César Pelli remembered his years with Saarinen (1954–1961) in similar terms: "We worked with a purpose. It was a wonderful time to be young.... Eero was not only a very gifted, creative architect, but he also had a great sense in selecting collaborators.... The main ingredient was enthusiasm. All in the firm felt we were part of an important enterprise."

The CBS Building, on Sixth Avenue and Fifty-second Street, was one of the last buildings Eero Saarinen worked on before he died. He was very excited about the 38-story tower, which was his first commercial building in New York and the only high-rise in his short career. It was a very exciting commission for me, as well.

For decades, there had been very little major construction in New York, what with the Depression and then World War II. And when, in the late 1940s and 1950s, new buildings finally were erected, they were fairly commonplace. Though there were good economic reasons for developers to embrace a formulaic approach to building, the final products all looked pretty much the same. I know because I had worked on many of them.

For the new CBS Building Saarinen and his client wanted something better, and they got it. It was my first headquarters, or first building specifically designed for just one company, and it presented me with my first major opportunity to do work that was imaginative and different.

Coincidentally, I had been involved with CBS several years earlier, when I worked on the studio of one of their evening broadcasters. The space occupied an area of only about 6,000 square feet, and was located deep inside the building for protection against street noise. I decided that if this commentator was going to speak to New York from a building in the heart of the city, he should always have reliable access to the air. I designed an emergency generator so he could continue broadcasting even in the unlikely event of a power failure.

Fast-forward a few years to "The Great Blackout of 1965" when New York was brought to a standstill during the evening rush hour, right in the middle of the commentator's program. He continued to broadcast without interruption, but during a commercial break, someone came in and said,"Hey, come look outside.

CBS broadcasting through the Great Blackout of 1965

Ductwork, wiring, and piping run inside the triangular piers of the CBS Building. The ductwork on the lower floors (seen below in the triangular column section at left) increases in size on the upper floors (compare right column section).

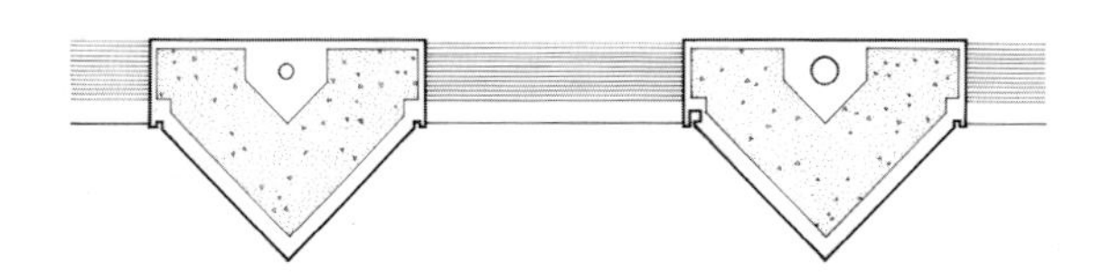

There's a power failure. The city is dark." CBS was subsequently commended in the press as "a prime example of preparedness."

The new CBS headquarters was among the first major buildings we worked on, and we were able to contribute some useful innovations. It is important to remember that the last major wave of construction in New York took place before World War II, which is to say before air-conditioning, before fire systems or sprinklers, before many of the building systems we have long taken for granted. Most of what we were doing at Cosentini was not experimental; it was new but it was based on sound engineering principles. But since none of it had actually been done before, and since the whole world was changing so quickly, it was easy to seem like a genius when I suggested an innovative system or process that actually worked.

It is also important to note that a fair amount of "sharing"—copying—went on from one engineer to another, from one job to the next so a lot of the new techniques I invented were borrowed by other builders and engineers. By the same token, when Edward Durell Stone's General Motors Building was completed on Fifth Avenue in 1968, I carefully studied the building (which I did not work on) and figured out what had gone right and what went wrong in its engineering. I took those lessons with me to Chicago for the Time-Life Building, which I worked on with Harry Weese.

Great post-war prosperity and technological advances led to lots of changes in the A&E industry. Old air conditioners, for example, used to be designed by manufacturers on their own without reference to other building systems or the engineers who designed them. The apparatus was cumbersome, complicated, and expensive to install, run, and maintain. Furthermore, most systems were manufactured in Europe, where the climate does not vary too radically from season to season, so they weren't ideally suited to North America's extremes. We first had to solve this problem in the late 1950s at Place Ville Marie, designed by Harry Cobb, as Montreal's temperatures range from minus ten degrees Fahrenheit in the winter to close to one hundred degrees in the summer.

The Carrier and Trane corporations developed into the leading American manufacturers of air-conditioning systems. I still remember how Trane executives would regularly come to my office to ask, "What are the problems? What should we be developing?" It was Trane that I persuaded to come up with a cooling device that I could use on separate floors for separate tenants. My competitors said, "That's the kind of air conditioner you put in a drug store," but such localized units became the industry standard; everyone now uses them.

It was only when the professional engineer stepped in that radical changes could be made in the climate control of large buildings. The new systems made deep floor plans possible,

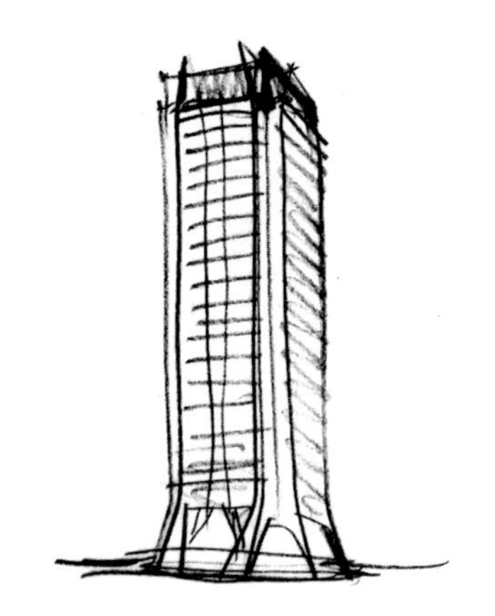

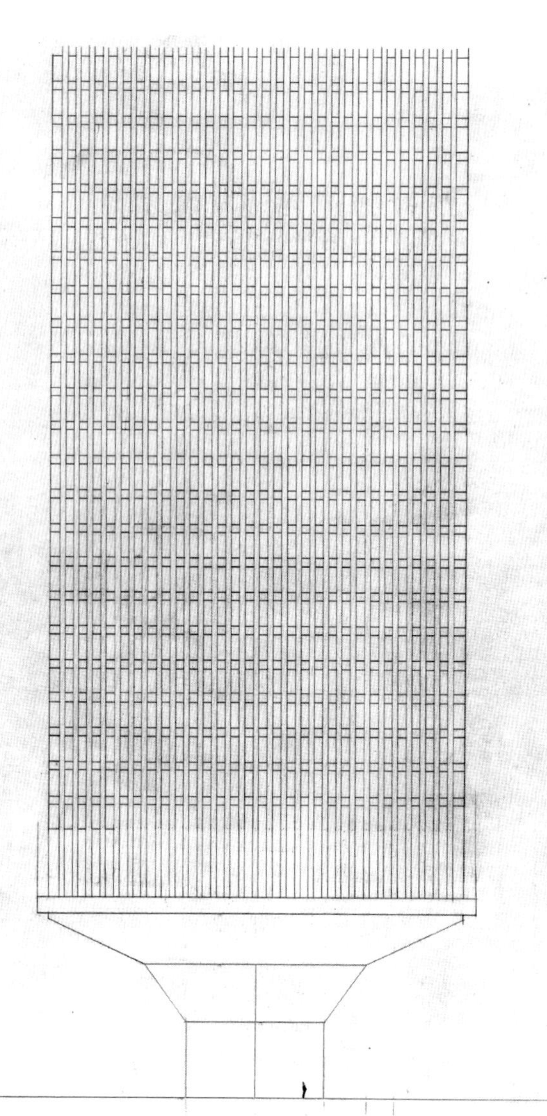

Among the early proposals for CBS was a "tree scheme" (above) which proved impossible due to site issues.

Other schemes (left) envisioned the building on skewed pilotis and set back in a landscaped plaza.

Opposite page: CBS Building viewed looking east across Sixth Avenue

since occupants no longer had to rely on exterior windows for air. In collaboration with architects and manufacturers, we designed small modular systems that were factory-assembled and self-contained. Because they took so little space and because they fed into specific areas, they eliminated inefficiency and saved money, and they were also extremely energy-efficient. Now, for the first time, you could track and control your own energy costs and condition your office space only when and where people worked in it instead of heating or cooling the entire building.

In Saarinen's original design for the CBS Building, huge columns along the perimeter were meant to be gathered together like a tree trunk at the base. I saw the model, and it was an amazing and powerful scheme. The problem was that the Sixth Avenue subway makes a turn right through the middle of the property. That made for terrific complications in construction and ultimately killed the "tree scheme," since the trunk would have intersected the subway tracks. Kevin Roche modified the design so that the columns, instead of being bundled, now run straight down the sides of building, 490 feet without setbacks.

The columns are triangular in shape and are a strong design element that seem to fuse into a solid wall when viewed on an angle. The columns were built of reinforced concrete and then clad in Canadian Black granite, which led to the building's nickname as "Black Rock." CBS was the first post-war skyscraper

"This thirty-eight story, freestanding tower set in its own shallow sunken plaza is unquestionably great architecture because it is original, consistent, boldly expressed and daring… This is a powerful building, whose angled piers thrust skyward with great energy, assertively expressing its dynamic structuralism in a manner that makes the Seagram Building seem almost dainty and frail. An important key to its cohesive expression is the equal division of its facades. Simplicity and focus are this building's bywords."

THE CITY REVIEW *Carter B. Horsley*

in New York to use an exterior load-bearing wall. Saarinen's goal was to break the mold of International Style steel and glass curtain walls, which were going up all over the city.

Inside the piers, which are hollow above the second floor, we ran all the ductwork and wiring and piping to feed each floor. The piers become more and more hollow as they rise, since loads diminish on the upper floors, while the air-conditioning ducts require increasingly more space to accommodate the growing number of floors feeding into them. To put it another way, the columns have several functions: aesthetic, structural, and mechanical.

CBS lobby (below); Florence Knoll and Eero Saarinen working on a chair binding (bottom)

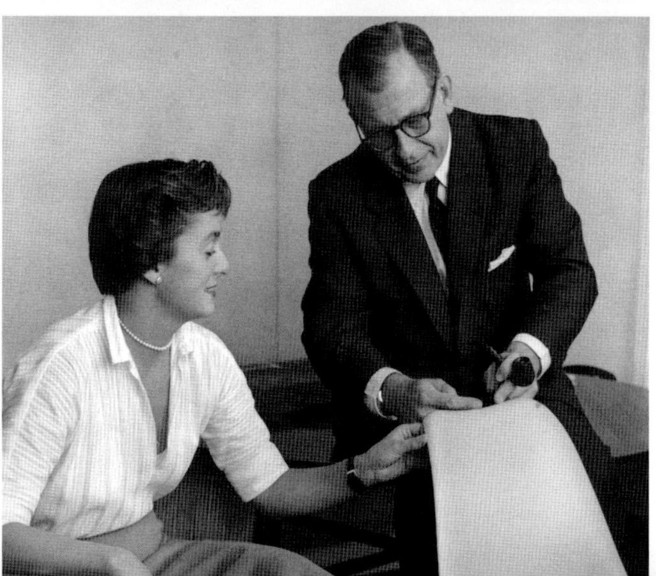

Five months before Saarinen died, he told the president of CBS that the new tower would be "the simplest skyscraper statement in New York," so efficient and simple, in fact, that it would "make the Seagram Building look gaudy.... Plans, structure, elevators, and the mechanical systems are all falling into place in a very organized way." To achieve that efficiency and simplicity, many complex feats of engineering were required.

The 5-foot-wide columns are each set five feet apart, and all the lighting and electrical outlets follow the same 5-foot measure. In other words, the whole building is modular so interior office partitions could be moved with ease. Terrific. But I had one big question: How are we going to bring in the huge refrigeration machines and other equipment? They wouldn't fit between those 5-foot openings.

The structural engineer's brilliant solution was to leave an open gap for the machinery and to fill in the space with fake, removable columns that look identical to their structural counterparts. The equipment required for the 800,000-square-foot building was such that it demanded a much greater floor-to-ceiling height in the mechanical floor than in the office floors. To minimize the visual impact, and to avoid the conspicuous banding that so often interrupts the facades of other buildings, we divided the mechanical floors into two, each servicing about twenty floors of offices. We located one at the second floor and the other at the top of the building, where they wouldn't be so noticeable.

Another fun experience at CBS was getting to work with the interior designer Florence Knoll on one of her most famous projects. Under her direction Knoll Associates designed all the offices, the furnishings, colors, everything but the travertine-lined lobby, which was done by Saarinen's office. One thing that sticks in my mind was a decorating problem that came up with the fire alarm system. City regulations required a red alarm for high visibility on the wall. But the interior designers felt the color was jarring and wanted the alarms to be white. Amazingly, we got the change approved by the New York City Fire Department. But later, when new interior designers were brought in, they wanted to add a little color—and the red alarm was restored.

Saarinen was the most unassuming great man I've ever met. He was pretty much the same even with his clients, most of whom were leaders of prominent industries and cultural institutions. Eero saw them as partners in the creation of a proud new post-war America. For the most part, he fulfilled their wishes, but since Eero believed that the architect was "not a mirror" and that he needed "the strength and urge to produce form, not compromise," he could also be insistent on his own ideas.

Eero Saarinen in his Tulip Chair

For the beautiful North Christian Church in Columbus Indiana, for instance, Eero took architecture in an adventurous new direction, extending design long past the point where the congregation wanted to stop. He wrote to the client: "It would be easy to say—as you would like me to—'Let's go ahead with [the design] as it is.' But against that I have perhaps a greater conscience because I know in my heart that it would not really be the best I can do."

I worked in Columbus from the late 1950s to the early 1980s, first with Saarinen and then

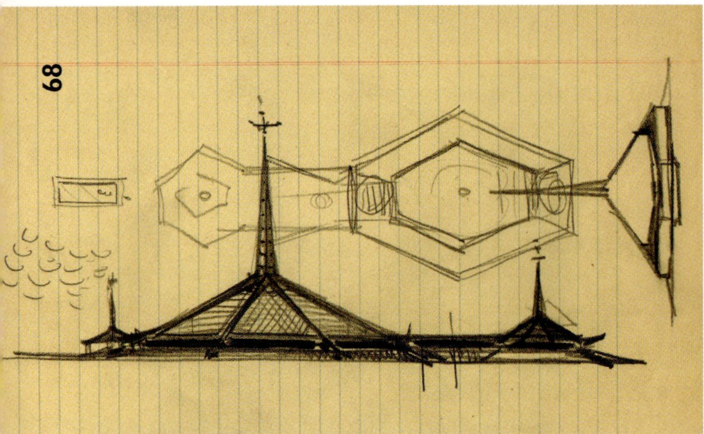

North Christian Church, Columbus, Indiana (1964)

with Kevin Roche, and Harry Weese. The enabling force behind their buildings and, in fact, behind most of the town's public buildings, was the Cummins Engine Company, one of the largest manufacturers of diesel engines in the word. The company's chairman was J. Irwin Miller, and he decided that the way to breathe new life into the town (and into Cummins Engine Company) was to attract top talent to Columbus by commissioning gifted architects to design new schools and public facilities.

Through Mr. Miller's efforts—and through his investment—Columbus became a model showcase of modern architecture with more than sixty landmark buildings, far more than in most other cities. As Paul Goldberger explained in the *New York Times*, "There is no other place in which a single philanthropist has placed so much faith in architecture as a means to civic improvement."

Columbus and its modern architecture became so distinguished that this small corn-belt town would earn the name "Athens of the Prairie." Ultimately it was to have four Saarinen-designed buildings, including Irwin Miller's own house, plus the First Christian Church, designed by Eliel Saarinen with Eero's help. The chaste rectangular church with its freestanding bell tower had introduced modern architecture to the town back in 1942, and was among the earliest modern religious buildings in this country.

From there on, the list of architects working in Columbus read like *Who's Who*. We did the engineering on several of the buildings, including Eero's North Christian Church and the Irwin Union Bank & Trust; the Cummins Engine Plant, and the Columbus Post Office, both by Kevin Roche, and Harry Weese's Hamilton Center Ice Arena. In a related job for a Cummins factory in the UK, we adapted the technology we had designed for Indiana, modifying the high-performance heating and air-conditioning systems of the Midwest for England's more temperate climate.

Like his buildings, Eero Saarinen himself was unique. Not only did he solicit his staff's ideas, he was one of the few architects I ever knew to give generous credit to the people he worked with. Eero supported and actively promoted his disciples. And because he was so collegial, and his working method was so inclusive, others could carry out his projects after he was gone. Saarinen gave his partners, Kevin Roche and John Dinkeloo, the freedom and authority they needed so that when he died, the firm continued. That kind of behavior is unusual among "star" architects.

Also unusual was Saarinen's willingness to listen to and take into account what we engineers needed to make mechanical systems function. He gave us the space we required without squeezing us into unworkable leftover corners. He was a thoughtful man who didn't merely sketch a design; he researched it and analyzed it from an aesthetic and functional point of view in order to fill the client's needs —and ours.

Saarinen's tragically short career ended with his unexpected death in 1961, at age fifty-one, barely two weeks after being diagnosed with a brain tumor. He left ten projects for Roche and Dinkeloo to complete, including Gateway Arch in St. Louis and the TWA Flight Center at Idlewild (now Kennedy International) airport. I admire both of these projects greatly although I did not work on either.

The arch, formally known as the Jefferson National Expansion Memorial, was Eero's favorite project and the first job he designed without his father. Actually, when the outcome of the design competition was announced, a congratulatory telegram was mistakenly sent

Irwin Union Bank & Trust, Columbus, Indiana (1954)

Gateway Arch, St. Louis, Missouri (Eero Saarinen, 1965)

to the more famous elder Saarinen. It wasn't until the next day that Eero learned he was the winner, not his father.

Eero worked closely on the arch with the gifted structural engineer Hannskarl Bandel, who helped determine the soaring shape and oversaw construction after Eero died. The arch is an ancient form like a great triumphal arch but for the modern age, and it was meant to symbolize, as Eero said, "the gateway to the West, the national expansion, and whatnot."

Like an inverted hanging chain, the catenary curve was designed so that its two triangular stainless steel legs, fashioned from narrowing interlocking segments, would join together at the top, at what would be called the keystone in a traditional arch. The joining was supposed to take place on October 28, 1965, at 10:00 a.m., but work actually began earlier in the morning.

The problem was that the sun had expanded the south leg of the arch by five inches so there wasn't enough room to insert the final piece. John Dinkeloo called the Fire Department and had them hose down the sunny side with thousands of gallons of cold water so that it would shrink to the correct size. By noon, as Vice President Hubert Humphrey patiently observed from a hovering helicopter, the final 10-ton segment was lowered into place. The legs were designed to join with an accuracy of just 1/64th of an inch.

Saarinen won the Gateway Arch commission in a two-part national design competition that took place in 1947–48 although construction did not start until 1963, two years after Eero had died. It remains one of the best known and most beautiful monuments in America, and at 75 feet higher than the Washington Monument, the 630-foot arch is also the tallest.

In the same way that the arch is a widely recognized symbol of St. Louis, Saarinen's TWA Flight Center was, and still is, a universal icon of aviation, a structural tour de force that expressed the thrill of flight in every detail. We in New York are lucky to have such a wonderful building.

I still remember how excited everyone was while its thin concrete shells were under construction and then again when the building finally opened in 1962. No one had ever seen anything like it. For years the terminal was the last word in glamorous air travel but in time, with airline deregulation and TWA's bankruptcy in the mid-1990s, the terminal was sold to American Airlines. The building's futuristic design proved difficult to update for new jumbo jets and new security requirements, so it was pretty much retired in 2001. There was talk of tearing down the magnificent structure, which would have been a travesty of the same order as destroying the old Penn Station. I can't understand how we can allow people who have no appreciation for architecture to remove a great building. Luckily, the plan to demolish the terminal was itself scrapped. JetBlue Airways bought and renovated Eero's "head house" and now uses it as a ceremonial entrance to the airline's much larger new terminal that partially encircles the original building along the back and sides. Components of Saarinen's original complex have been demolished, but what a relief that most of the soaring building is still with us.

"We wanted to design a building that would express the drama and specialness and excitement of travel. Thus we wanted the architecture to reveal the terminal, not as a static, enclosed place, but as a place of movement and transition. We wanted an uplift."

Eero Saarinen

TWA Terminal, New York, NY (Eero Saarinen, 1962)

Kevin Roche installing curtain wall on a Ford Foundation model

Of all the buildings I have been involved with over the past six decades, my favorite is the Ford Foundation, on East Forty-second Street in New York. It was designed by Kevin Roche after Saarinen's death.

The 12-story headquarters, completed in 1968, was the first full atrium building in the United States. It was the inspiration for the soaring atrium lobbies that became trademark features of Hyatt hotels, and the original model for public winter gardens in shopping malls and office buildings across the country. The Ford Foundation basically turned the traditional office building inside out, departing from the cool abstraction of the contemporary International Style to develop a sense of community and humanism in the workplace. Kevin Roche explained that he wanted to bring some interest to the lives of the people who spend eight hours a day in the "dulling, dulling, dulling" office work of most businesses. "There's no relief.... What can you do to make these environments better for people?"

Instead of looking out the window at nearby buildings crowded together across the street, Ford Foundation employees look through glass walls into a magical 1/3-acre public garden, almost like a jungle, in the middle of a private corporation, in the middle of the hectic city. Ada Louise Huxtable, architecture critic for the *New York Times*, called it "one of the most romantic environments ever devised by corporate man."

The Ford Foundation was one of our first green buildings, although, right around the same time, we did the GSIS (Government Service Insurance System) building in the Philippines designed by Howard Elkus, a principal at The Architects Collaborative (TAC). That building had a series of set-back roofs on which grass was grown. The tropical sun did not heat the building but instead, its energy was used to grow the rooftop lawns that, in turn, helped the building to stay cool. It is not that we thought about green construction in those days, but then, as now, we were concerned to make building occupants as comfortable as possible.

The Ford Foundation atrium rises 170 feet to a large skylight. We came up with a plan to discharge the used air from the offices into the building's atrium, where its CO_2 nourishes the plants. Then, cleaned by the greenery, the air is brought down into the basement. From there, we recycle it to the cooling tower, from which it feeds the air-conditioning. After that, the air is discharged.

Ford Foundation, New York (Roche Dinkeloo, 1968). View across 42nd Street from Tudor City

Interior, Ford Foundation

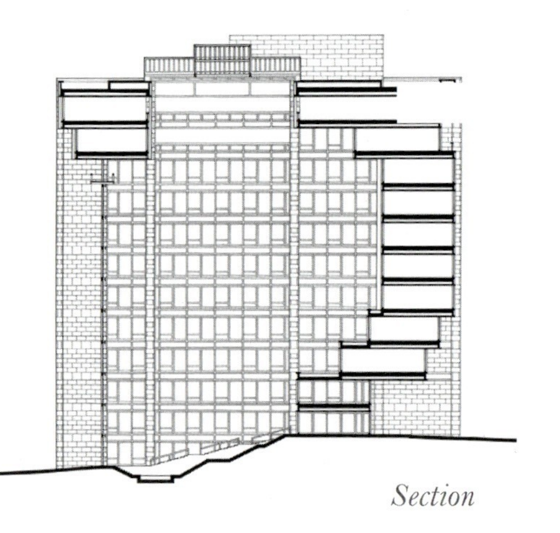

Section

"More than four decades after its completion, the Foundation is still a remarkably prescient piece of architecture. It excels in several areas where many architects continue to struggle: how to integrate natural light and decent views into the workplace; how to provide privacy to workers without sacrificing a feeling of community (or sequestering them in bland cubicles); and how to create a daring, iconic form that is a good neighbor and a true contribution to the city."

Mason Currey

"Rediscovered Masterpiece," *Metropolis* Magazine, December 2008

The Ford Foundation wanted a unique identity and, of course, they had the money to pay for it. Over twenty years later, when I. M. Pei's John F. Kennedy Library opened in Boston in 1979, we used the same technology in the atrium of that building (*see Chapter 5*). But it is only now that the nation's builders are beginning to appreciate, and are trying to redevelop, this kind of technology. In fact, if the Ford Foundation were built this year it would get a LEED Platinum rating, the highest category of the contemporary Leadership in Energy and Environmental Design program.

The Ford Foundation is a reinforced-concrete structure clad in pink-brown granite. But it makes significant use of steel, particularly in its great Cor-Ten steel spanning members. Most steel comes out of the factory with an oil coating that is extruded with the steel. In ordinary construction the oil is removed and the steel is encased in concrete. Cor-Ten involves a different process and a different finish. When the steel is extruded, it is pre-weathered to create a thin rusted skin that protects the inner steel from further corrosion by water and other atmospheric elements. The exposed Cor-Ten beams give the Ford Foundation a rich, warm brown appearance.

Steel plays an important role behind the scenes as well. The design called for huge spans of glass, about eighty feet long, facing south and east. There were to be no columns. To support the weight of so much glass, a 10-foot-deep concrete beam would have been required. Instead, Kevin Roche suggested to the structural engineers that they use a 2-foot steel beam, encase it in concrete, and then clad it with granite to match the rest of the building. That's exactly what they did. And we fit the steel beams with heating to keep the windows from icing. The Ford Foundation is a good example of the need for architects and engineers to understand the requirements they each have in executing a design and in working together to make a building work for the user.

Radiators encircle the horizontal framing members of the window walls, while air blows down from the roof. The system only operates in the winter, when the outside temperature falls below 25 degrees Fahrenheit. In summer, despite the large greenhouse enclosure, the load to cool the garden is still less than in the offices, where lights and office equipment emit heat. Meanwhile, in the garden, irrigation and feeding is supplied by a network of underground pipes to eleven different planting zones.

The Ford Foundation was an instant success, credited by *Architectural Record* as having created "a new kind of urban space." The architecture critic Paul Goldberger agreed in *The New Yorker* that nothing before had quite the same quality, "And not that much after, either." The building became a New York City landmark just as soon as it turned thirty and became eligible, and it received the AIA's 25-Year Award in 2008. As Mason Currey recently exclaimed in *Metropolis* magazine: "Visiting the Foundation today is still a unique and thrilling experience, one of those New York moments that should not be missed."

We worked with Kevin Roche on the Knights of Columbus headquarters, which opened in 1969 in New Haven, Connecticut. It is an early concrete building and is a wonderful example of the partnership between architecture and mechanical engineering. The imposing 23-story building, designed like a fortress for the Knights, is amazing-looking and at the same time it's a technical marvel. The building's corners are anchored by four massive cylindrical piers constructed of continuously poured concrete and sheathed in dark brown tile.

As in the Ford Foundation, the building uses exposed pre-rusted Cor-Ten steel, in this case as great 80-foot-long girders that span from corner to corner. Inside the cylinders are service towers with utilities, lavatories, and fire stairs; the elevators are located in the central core. In a seamless meshing of disciplines, we used the structural frame of the building to carry the air required for heating and cooling the interior.

In terms of energy conservation, the Knights of Columbus building was way before its time. Its windows are set back about two feet from the structure's edge with louvers that prevent the sun from streaming in and heating up the interior. In fact, back in the 1960s and 1970s, long before today's environmental movement began, we at Cosentini, together with Saarinen and Roche and Dinkeloo, were constructing high-efficiency LEED Platinum buildings before the LEED concept existed.

KEVIN ROCHE: *During a recent talk at the Ford Foundation, I saw Marvin in the audience and publicly acknowledged the very important role he had played in the building. The question came up, as it had so many times in the past, about the garden and the use of the space, and how it all worked in terms of growing materials and air movement. Marvin was far ahead of his time regarding energy conservation, returning all the air from the offices to the atrium and then recirculating it. This was a very advanced idea back in the 1960s.*

Marvin was just wonderful to work with, very smart, a very good engineer, and a delightful personality. He was always creative, with lots of good ideas and suggestions, not just slogging along. I used to look forward to meetings with him.

Marvin ran a tight ship and his people always performed very well. There were no aftershocks, nothing was left undone. It was just high-quality performance. But the best part was Marvin himself—he's a delightful person.

In a remarkably selfless way, Marvin never put his name on the business. But he didn't need to because everybody knew. The firm name might be Cosentini, but it was always Marvin Mass.

Knights of Columbus headquarters, New Haven, CT (Roche Dinkeloo, 1969)

Left to right: I. M. Pei, Henry N. Cobb, Araldo Cossutta, James Ingo Freed

PEI, COBB, COSSUTTA, FREED

I. M. Pei is one of the most gentlemanly and considerate architects I have worked with. Once, talking about Louis Kahn, I. M. said of his own temperament, "I'm a little more patient with clients. If a client doesn't agree with me, I let it go and come back another day." Patience and endurance are part of I. M.'s 5,000-year-old cultural heritage.

I. M. Pei is known for the sculptural geometry of his Modernist designs, with their advanced technologies, sharp angles, and bold abstract forms. From my point of view, his buildings are extremely adaptable. Even before his schemes were on paper I. M. would sit with us and get our input so that all the electrical and mechanical systems were fully integrated into his design. I don't know why other architects didn't do that. Louis Kahn, for example, would adapt his plans to accommodate the necessary systems after the fact.

Pei was born in China in 1917, and received his bachelor's degree from MIT in 1940. After a travel fellowship, he enrolled in the Harvard Graduate School of Design to study under Walter Gropius. In 1948 Pei was recruited to be an "idea man" at the Webb & Knapp real estate company for the larger-than-life developer William Zeckendorf. I. M. worked as Bill's

the AIA, which censured him for being a "captive architect"). Pei gave Zeckendorf a new respectability and put him on the map with designs unlike those that any other developer was doing. There's no question about it. When I. M. was there, all sins were forgiven.

When Pei went out on his own in 1960 he kept his same rooftop offices in the building adjacent to the penthouse headquarters he designed for Zeckendorf, at 383–385 Madison Avenue. You couldn't go directly back and forth between the two offices because there was a little valley in between, but you could walk on the parapet past the elevator housing for access, as the architects often did. The two men parted good friends and continued to work together on projects like Kips Bay Plaza in New York before Zeckendorf's empire collapsed in the late 1960s. I. M. kept in touch until the final days of Bill's life.

ALDO COSSUTTA

One of Pei's early partners was Araldo A. Cossutta, a Yugoslavia–born, École des Beaux-Arts–trained architect who joined the firm in 1956. Having worked in Le Corbusier's office when the landmark Unité d'Habitation was in progress, Aldo was an important force in the change away from the steel, glass, and aluminum favored by the International Style toward the extensive use of cast-in-place concrete. Beginning with Kips Bay (1957–62), Pei's firm championed precision poured-in-place concrete buildings for years to come.

Aldo is a meticulous and persistent architect who is well known for constantly making changes to his drawings, right down to the wire, in endless search for a better solution. He was also a shrewd businessman. He left Pei's office for independent practice in 1973 with a big job in hand, the 42-story Crédit Lyonnais Tower in Lyon, France. In addition to his commercial ventures, Aldo designed his own beautiful house on Martha's Vineyard.

We were involved with Aldo on a major project commissioned by the Thompson family, who were owners of the 7-Eleven chain of convenience stores. The 42-story Towers at Cityplace were to be the flagship headquarters of a planned 130-acre mixed-use complex in central Dallas. To feed this huge project we designed a central chilling plant with ten thousand tons of refrigeration. It was the largest chilled water storage system in the country, with 10 million gallons of water (the equivalent of 15 Olympic-sized swimming pools). The plant was constructed in Phase I, but after the first tower was completed in 1988, the project stalled, with the result that we had overdesigned the central plant. Eventually, residential buildings were constructed and serviced by our system, so it all worked out in the end.

Our design was one of the first systems to use automatic controls for central cooling equipment; previously each had to be turned on manually, one at a time. In this facility, six chillers and also six cooling towers were manifolded into one system to serve the entire project. We designed the system to turn on the various functions in sequence and to use the equipment alternately, rather than to leave any of it dormant for long periods of time. After the building was up and running I met with the chief engineer to see how things were working. He assured me everything was fine, but when I asked him to start pump number three, pump number two went on. The same thing happened with the chillers. It turned out that they had been hooked up incorrectly, but everything finally got straightened out. I was glad I'd made my visit.

For a long time, Pei's firm was known as I. M. Pei & Partners, but its name was changed in 1989 to Pei Cobb Freed & Partners in recognition of the contributions of I. M.'s longtime design partners Henry "Harry" Cobb and James Ingo Freed. Pei officially retired in 1990, but over the next two decades he frequently collaborated with his sons Chien Chung "Didi" Pei and Li Chung "Sandi" Pei, who left PCF to found their own firm, Pei Partnership Architects, in 1992.

When China was first emerging from strict Communism in the 1970s, I. M. was asked to design a new hotel for the waves of Western tourists expected. The government wanted a big modern building, but I. M. resisted because he didn't want to intrude on the open skies of the Forbidden City. With even a single skyscraper in the center of town, China would lose her precious architectural heritage forever. Instead, I. M. went about thirty miles outside the city to the old imperial hunting grounds where he knew there would be no pressure to build tall. The remote site was occupied by a bare-bulb hotel in such bad condition that I. M. had it demolished. All that was left were the original foundation walls and the beautiful

Cityplace, Dallas (Araldo Cossutta, 1988) and its central mechanical plant and chilled water storage system

Fragrant Hill Hotel, Beijing (I. M. Pei, 1982)

traditional gardens with their hundreds-of-years-old trees.

This was an intensely personal project for I. M., who hadn't been back to China since he left as a teenager in 1934. He wanted to develop an architectural language to help local architects in moving forward. After the work was finished, one big problem remained: what to call the new hotel? There was a lot of discussion about an appropriate title, until one day a government official called to say that he'd found a solution. "I've just heard that in the United States there's something called the Quality Inn. That's what we should use." He didn't understand that quality is relative and doesn't necessarily mean deluxe; he did not seem to appreciate the distinction between I. M.'s unique design and an affordable national hotel chain. In the end, Fragrant Hill Hotel took the name of a nearby park.

I worked with Pei on the John Fitzgerald Kennedy Presidential Library and Museum in greater Boston, one of the most important buildings of the day. After watching a brief film about JFK's life and career, visitors proceed to a series of underground exhibition spaces with no natural light or views. From this darkened setting they enter into a great skylit atrium, "a restful place, where they can linger, look at the view, and reflect on what they have seen. In the silence of that high, light-drenched place," I. M. continued, "visitors are alone with their thoughts. And in the reflective mood that the architecture seeks to engender, they may find themselves thinking of John F. Kennedy in a different way. In the skyline of his city, in the distant horizons toward which he led us, in the canopy of space into which he launched us, visitors may experience revived hope and promise for the future."

It was not an easy project to get built. Pei won the commission for the library in 1964 and overnight he became an international sensation. But years of strife followed as the United

States struggled with the civil rights movement, the unpopular war in Vietnam, student unrest, the critical evaluation of JFK and his administration, and, in practical terms for the library, the steady depletion of its construction budget to cover growing administrative costs. Together these forces dampened the zeal to build a memorial for the assassinated president.

It was strange for a building of such national importance, but one of the biggest problems was finding the right site. The process dragged on for eleven long years as multiple possible locations were investigated and then rejected. Ultimately the library was shunted from a prime site near Harvard Square to end up in Dorchester, close to the childhood home of Rose Kennedy, where JFK began his public life in the House of Representatives. Groundbreaking took place in 1977, and—full speed ahead—the building opened to the public two years later.

Kennedy Library included one of the country's early glass-enclosed atriums; it looks out over the water and downtown Boston. My problem was to make the all-glass box comfortable in New England's temperature extremes without the mechanical systems showing or disturbing the silence of this powerful 110-foot-high memorial space. My experience with the reuse of spill air at the Ford Foundation was very pertinent.

The library presented yet another unique and curious challenge. Because the building was situated on landfill, a former garbage dump, waste material, just under the surface, was producing methane gas, a normal byproduct of decomposing matter. For protection, an elaborate ventilating and drainage system had to be designed. The building was specially sealed and situated on a flat slab fifteen feet above the ground to bypass a major sewer pipe. Its split-level organization is the happy result of this grade change.

Section (below) and Main Hall (right), John F. Kennedy Library, Boston (I. M. Pei, 1979)

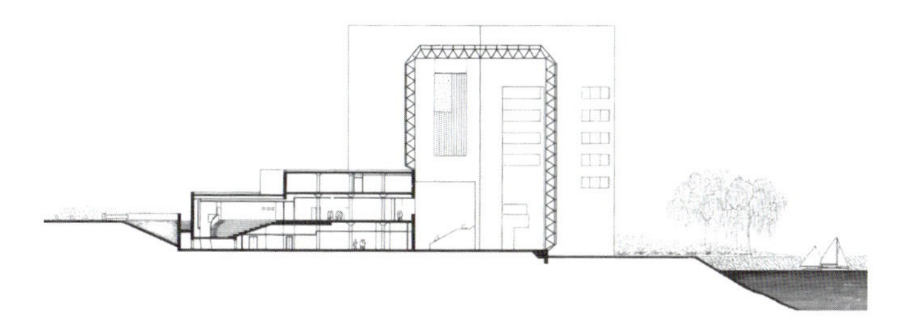

HENRY N. COBB

I. M. Pei's partner Harry Cobb was responsible for the John Hancock Tower in historic Copley Square, the heart of Boston's Back Bay. The 60-story building was severely criticized because it was feared that it would overpower surrounding architectural masterworks, such as the Copley Plaza Hotel, H. H. Richardson's Trinity Church, and McKim, Mead & White's Public Library. Nobody makes such comments anymore. Because the glass facade is so flat and highly reflective, the older buildings are projected onto the tower's surface and can be enjoyed again in their mirror image. And since the minimalist building is so neat and sleek, it adds significantly to the neighborhood and to the Boston skyline rather than detracting from them or seeming intrusive.

In 1973, while the tower was still under construction, its large 500-pound windows began to crack. In the following weeks the number of window failures grew, generating a lot of negative publicity and alarm, and also genuine concern, because no one really understood what the problem was or, more important, how to fix it.

We were the engineers on the project, and the first diagnosis was that the failure must have been my fault because I pumped in too much air and overpressurized the building. I humbly pointed out that the cracks appeared before the air systems were even working.

The true cause was kept secret for many years, particularly since all the signatories were, and still are, legally bound to silence. Even so, enough data has been compiled over the years to pinpoint the problem to the windows themselves. Thermopane windows had been used before but never in such a prominent building as the 790-foot-high Hancock Tower, which, even now, 35 years later, is still the tallest in New England.

The windows consisted of two glass panes bonded together by a brass spacer, leaving an air space in between; the sandwich-like assembly was then set into a metal window frame. To cut glare and solar buildup, the outer glass pane was given a silvery chromium coating on its inner, or number 2, surface. This accounted for the building's mirrorlike reflectivity.

The problem was that the bond was too rigid to allow the expansion and contraction that routinely happens with temperature change, nor did it allow the joints to vibrate in the wind. The bond fatigued and then failed, pulling a little chip of glass with it. This weakened the glass and set in motion the failure of the entire window. Ultimately the glass company paid to replace the windows, all 10,344 of them, with single-pane tempered glass and flexible bonds to accommodate expansion and contraction. (The manufacturer also quietly replaced the windows in all of the other buildings that experienced similar problems.)

John Hancock Tower, Boston, MA (Henry N. Cobb/I. M. Pei & Partners, 1976)

Whatever its problems had been, the Hancock Tower went on to win the hearts of professionals and laymen alike, and is among the purest and most admired skyscrapers anywhere. In 1977, the year after completion, it received a National Honor Award from the American Institute of Architects. Local architects and historians subsequently voted it the third-best building in Boston after Trinity Church and the Boston Public Library, the very neighbors the glass prism was designed to complement. That Hancock's allure continues was recently made clear when the AIA conferred on it the Twenty-five Year Award for enduring function and aesthetic achievement (2011).

499 Park Avenue, New York
(James Ingo Freed/I. M. Pei & Partners, 1981)

As a result of Hancock's window problem, I. M. Pei & Partners knew more about glass curtain wall than any other architectural firm in the country, maybe the world. Pei put that knowledge to good use when he later designed the Louvre Pyramid in Paris.

I. M. is one of my good friends, and I have worked with his firm on many projects, though usually with Harry Cobb or Jim Freed. Those men were responsible for many of the firm's buildings, but I'm afraid they haven't had the public recognition they deserve. To some degree the trials and triumph of Hancock Tower changed this for Harry. For Jim, recognition came when his extraordinary United States Holocaust Memorial Museum opened in Washington, D.C., in 1993.

JAMES INGO FREED

I had known Jim Freed for a long time before we worked together on the Holocaust Museum. In fact, in the 1960s, we each lived with our families in Kips Bay, the landmark concrete housing complex on which Aldo, Jim, and I all worked with I. M. in New York. I had also worked with Jim on several other projects, including his elegant skeletal tower at 88 Pine Street in New York, and also 499 Park Avenue. This dark glass tower, completed in 1984, was a landmark building for us, as this was the first time we used packaged air-conditioning units with a "free cooling cycle." We installed a cooling tower on the roof to perform double duty: it cools the air before it reaches the compressors in the individual units and then it also removes the heat from the compressors.

Twenty percent of air circulating through a building must be from the outside. In the summer that means you are bringing air inside that is 90° to 100° Fahrenheit. With the use of the cooling tower, we were able to lower that temperature by ten or fifteen degrees before it even got to the compressors, thereby saving lots of energy.

To put it another way, we've all seen air-conditioning coils in our home units. They're powered by electricity, as we well know from the many warnings issued to cut back on air-conditioning usage. Employing another approach to the problem, we designed an energy-saving scheme whereby the liquid in the coils at 499 Park is primarily cooled by evaporation, which uses much less energy.

But back to the Holocaust Museum. James Ingo Freed was born in Essen, Germany, in 1930, just as Hitler was consolidating power. At age nine, he was sent from Germany with his four-year-old sister, just the two of them traveling alone, to live with relatives in Chicago; eventually their parents joined them.

Jim was not an observant Jew, and it was not until his work on the museum that he ever got in touch with his heritage. It was a difficult journey with profound personal consequences. The Holocaust Museum was also the ultimate professional challenge for this Mies-trained architect, as it required him to step away from the clean, objective forms of his past and to delve instead into the realm of history, the subconscious, the irrational, the unthinkable.

Jim wanted to somehow express the Holocaust architecturally but without literal reference to places or events, or worse yet, "Disneyfication," to use his word. His brilliant solution was to let the building speak to each person's own experience in a raw and visceral way. As Jim explained, "The intent of the building is to be a resonator of your imagery, of your own memory." It is one of the most powerful and emotional buildings in the world.

During the course of my work on the Museum, I was most deeply moved when they brought in the actual boxcar that had transported people to the concentration camps. The mere sight of the windowless car that had carried so much misery and suffering was for me a defining moment. It took history out of the past and made the whole experience very personal. Though my parents were in the United States during that period and did not suffer directly, still, who would not be brought to tears by the barbed wire behind which starving people were kept, or by the very train into which they were herded?

As to my work, the trickiest part was dealing with the objects. We are not talking here about precious oil paintings or fragile sculpture but about the shoes of the victims, the women's skirts and dresses that were removed before they were gassed, the tower of thousands of irreplaceable photographs showing people, young and old, living the fullness of daily life before their entire village was exterminated in a single 24-hour period. Temperature and humidity had to be maintained in areas where such artifacts were displayed.

One of the challenges in museum work is that the public is usually in the building only from 10:00 a.m. to 5:00 p.m. while the artifacts are there twenty-four hours a day, seven days a week. Therefore, we designed a dual system: I made sure there would be plenty of fresh air in the museum during the day, but to save energy I cut the flow down after hours. We maintain the temperature and humidity, of course, but we recirculate the air rather than bringing in a new supply.

Some of the spaces in the museum, like the 7,500-square-foot Hall of Witness, are huge. When people ask me how we handle the immense height of such spaces I explain that we supply air and heat and cooling to the areas that people occupy and could not care less what happens forty feet above their heads. Therefore, the extremely high ceilings in the building's main public spaces did not concern us.

The only exceptions were the narrow glass bridges that run outside the building to span the museum's skylit central space. With their glass walls and floor, the bridges seem to float precariously in the air. Freed deliberately designed them to be a little frightening to cross in order to evoke some of the fears and discomforts endured by Holocaust victims.

We designed the mechanical systems to maintain the year-round humidity required for the museum's precious artifacts, but because the glass bridges are above the building, rather than inside, winter condensation posed a special challenge.

On every project I'd ever done with Jim Freed, he'd say, "Marvin, I don't want to see, hear, or feel your designs because if I do, your systems aren't working right." The Holocaust Museum was different because now, at least in the Hall of Witness, the building's main public space, our work was to be on view, since it reinforced the industrial aesthetic that Jim was after. It was a challenge, and also a great honor, to work with Jim on this brilliant building.

Left: Rail car en route from Poland to the museum

Hall of Witness, United States Holocaust Memorial Museum, Washington, D.C. (James Ingo Freed/Pei Cobb Freed & Partners, 1993)

Left to right: Philip Johnson and John Burgee; John Burgee; Alan Ritchie; Philip Johnson

JOHNSON, BURGEE, RITCHIE

Philip Cortelyou Johnson came from an old American family. An early forebear had laid out the first town plan of New Amsterdam for Peter Stuyvesant back in 1660, so Philip came to his profession through genes as well as training. Born to a wealthy family in Cleveland, Ohio, in 1906, he studied philosophy at Harvard, and then traveled extensively in Europe, where, in 1928, he came under the profound influence of Mies van der Rohe. Johnson was to be instrumental in Mies's emigration to the U.S. several years later.

I begin with this because, shockingly, Philip Johnson was a Nazi enthusiast in the 1930s and spent time on the front lines as a correspondent witnessing what he called "the stirring spectacle" of Warsaw's destruction. In one of the many about-faces that characterized his life, Philip left Germany in 1933, returned to Harvard to study architecture, and a year later, at age 34, joined the U.S. Army. (The other soldiers called him "grandpa.")

Johnson later said of his fascist leanings, "I have no excuse for such unbelievable stupidity.... I don't know how you expiate guilt." I worked with Philip for almost forty years and I never experienced even the slightest hint of anti-Jewish behavior from him. In fact, some

HONORING PHILIP JOHNSON

Left: Marvin Mass in his office, 2011

Below: The Mary Buckley Endowed Scholarship Dinner for the Pratt Institute at the Sony Club, New York, 1993

years back, we wanted to honor Philip as an industry leader in the architectural division of Israel Bonds. But after announcing his acceptance to the committees, there was some dissension on the board and it fell to me to explain the reversal. Philip said instantly: "I totally understand. I am sorry for what I did when I was a kid. Marvin, don't feel bad. I'm not offended at not being honored. I made a terrible mistake and I have to live with it."

Philip was notorious for comments that would have been better left unsaid but which, once voiced, became unforgettable like, "There's a little whore in all of us." He enjoyed being an *enfant terrible*, although at age 85 he admitted that he was getting a little old to be an *enfant*. Philip Johnson was the most clever and influential critic, reputation maker, and trendsetter in modern American architecture. He introduced the International Style to the United States with a watershed book and exhibition at the Museum of Modern Art in 1932. Fifty years later, he made an about-face and embraced ornament and historicism at the AT&T (now the Sony) Building in New York. The truth is that Philip's buildings were Modernist, Postmodern, Constructivist, Deconstructivist, or whatever else the client wanted to see.

Philip was a very anxious, nervous man. A couple of days after John Dinkeloo died, Philip, his partner John Burgee, and I were on our way to the funeral. John was driving and Philip was sitting in the back seat. He kept saying, "Go faster. Pass that car. What's taking so long? Go faster, faster." I remember this jittery scene vividly because Philip's urgency was so at odds with the timelessness of death. And speaking of funerals and Philip: I remember him at Gordon Bunshaft's memorial service. "Gordon and I never agreed on anything," he said, "and we never liked each other. I never really thought very much of Gordon but now that he's dead, let me tell you, he was a GREAT architect."

Philip Johnson began his life in architecture at the Museum of Modern Art, where at age 26 he established the first Department of Architecture and Design at any museum in the United States. Aside from a few small buildings, including the Glass House, which Philip built for himself in 1949, he had never done a major work, so when Phyllis Lambert sought MoMA's guidance in selecting an architect for the Seagram Building, Philip helped steer the commission to Mies (over Frank Lloyd Wright, Le Corbusier, Marcel Breuer, I. M. Pei, and Eero Saarinen). Philip became associate architect on the 38-story tower, and was personally responsible for the design of the sensational Four Seasons restaurant.

The Seagram Building established Johnson as a practicing architect, but his most famous buildings date from his partnership with John Burgee (1967–1991). During this very productive period, we worked together on the Boston Public Library (1972); IDS Center, Minneapolis (1973); Crystal Cathedral in Garden Grove, California (1978–80); the AT&T Building in New York (1980–84) and also PPG Place, in Pittsburgh (1981–84). In addition we worked

on the elliptical "Lipstick Building" at 53rd and Third (1986), where Johnson and Burgee located their office in 1986.

One of the more interesting buildings we did with the firm was the twin-tower Puerta de Europa in Madrid (1989–95), which were the first inclined skyscrapers in the world. A major subway exchange at the busy Plaza de Castilla required the two buildings to be set far back from the street, but to create a compositional whole, they were made to slope toward each other at a 15-degree angle. Their convergence over the Paseo de la Castellana, Madrid's most important boulevard, provided a monumental entrance into the city's financial district and, more than that, a symbolic gateway into all of Europe. For clear identification from the air, the western rooftop heliport is painted blue, while the one on the east is painted red.

After Johnson and Burgee parted ways in 1991, Alan Ritchie, who had joined the firm back in 1969, formed a partnership with Johnson. Philip was the dreamer and Alan was in charge of running the office and getting everything done. When Philip got sick, Alan assumed responsibility for producing design and construction details for all of the projects, and when Philip died, Alan continued the firm's work and actually resumed work that had previously stalled. I lived through a lot of this with Alan and in the course of it, we became very good friends.

Top: Puerta de Europa, Madrid (Philip Johnson and John Burgee, 1989–1996)

Bottom: 53rd at Third (the Lipstick Building), New York (John Burgee Architects with Philip Johnson, 1986)

Among the projects that Philip Johnson/Alan Ritchie Architects did were the Trump International Hotel and Tower at Columbus Circle in New York, and the Cathedral of Hope in Dallas (1995). The world's largest all-inclusive Christian church was the project by which Philip intended to be memorialized. The cathedral remains to be built, but in 2005, just months before Philip died, he completed sketches for the 225-seat Interfaith Peace Chapel. It was built in the kind of "blob" architecture that Johnson experimented with in his later years. The chapel was dedicated in November 2010.

As we have seen in Chapter Three, in 1939 Edward Durell Stone in association with Philip L. Goodwin designed the original Museum of Modern Art on West Fifty-third Street in New York. I, of course, did not work on the original building, but from that time up until the latest addition by Yoshio Taniguchi in 2004, the building seems to have been a magnet for me.

In the context of the Holocaust Museum, I mentioned an engineering challenge specific to art museums: air must circulate not just in the public spaces but also over the object on display. When I first confronted this challenge I looked to the great old museums of Europe, where historically air came inside from above and left through grills along the perimeter of the galleries. We improved upon this model by installing floor vents below the individual artworks so that the air passes right over the art on its way to being exhausted.

In the past, museum management had maintained that, to preserve precious works of art, air-conditioning systems had to function at full capacity twenty-four hours a day. The energy consumption was enormous. I figured that there had to be a more efficient solution. I laughingly say that we process the meat, meaning that we typically condition the air in areas people occupy but don't bother about the spaces high above their heads. By extension, if there are no people to make comfortable, let's condition the air as required. So, when the museum is closed, we reduce the amount of refrigeration while still ensuring circulation over the art. In other words, we achieve considerable savings without any loss in quality.

Museum of Modern Art, original building (Edward Durell Stone and Philip Goodwin, 1939) and MoMA residential tower (César Pelli, 1984)

AT&T (now Sony) Building, New York (Philip Johnson and John Burgee, 1988)

One of the more memorable buildings that I worked on with Johnson & Burgee was the AT&T headquarters on Madison Avenue in New York. It was one of the few buildings we designed exclusively to meet the particular needs of a single occupant. Of course, what happened was that roughly twenty-five years later, following corporate divestiture and downsizing, AT&T no longer needed this highly customized building, and no other company had the same specialized requirements, meaning nobody else wanted it. Certain floors had no load requirements; one floor had been planned exclusively for telephone systems, another for a private dining room, and so on.

Eventually Sony leased and later bought the building but had to make major changes, like bringing in more power, and also transforming the original atrium into retail space. The beautiful winged "Spirit of Telecommunications" sculpture, 22 feet high and covered in gold leaf, which used to command the building's 7-story lobby, took flight with AT&T's move to the suburbs of New Jersey.

Despite all the criticism, most of it directed at the building's Chippendale roof, I thought the general design of the AT&T Building was terrific. Philip did three or four other Postmodern buildings with elaborate ornamental crowns right around the same time. I remember I was working with him on International Place in Boston's Financial District, overlooking the harbor. Philip had made a model and I was with him at the client presentation. When he put the flat-topped buildings on the table, the client asked what the top was going to look like. "That's it," Philip replied. "That's it?" "Yeah," said Philip, "I'm finished with the other shit."

This story and lots of other prickly tales aside, Philip was, in fact, the epitome of the accommodating architect. He gave the client what he wanted. If someone said, "That model for the building looks too tall," Johnson would answer, "Okay, I'll make it shorter." Or, "Philip, what's that hole in the side of the building?" Johnson wouldn't try to defend it; he'd say, "Let's take it out."

Johnson was an avowed atheist but he liked to work on churches (he did at least five) because of their greater creative possibilities over "layers of office cubicles." One of his greatest commissions came in the 1970s when Philip was asked to design the Crystal Cathedral in Garden Grove, California, by Dr. Robert H. Schuller, a charismatic Hollywood televangelist who launched his ministry by preaching from the snack bar roof at a drive-in theater.

Schuller had hired Richard Neutra to design a permanent drive-in/walk-in sanctuary in 1958 and, recognizing the power of television, began on-air services twelve years later. Schuller's "Hour of Power" broadcasts were so successful that by 1975 a new megachurch was needed to accommodate his growing congregation plus all the cameras and equipment required for his international television viewing audience of some twenty million people.

When Dr. Schuller first talked to me, he said he wanted "a building that God would heat and cool." Philip's design called for an 86,000-square-foot cathedral that would seat nearly three thousand parishioners. To add to the challenge, the space was to rise to a height of 128 feet and, because Schuller wanted to recall the outdoor beginnings of his ministry, it would be enclosed entirely by glass. More than ten thousand panes were assembled in a web of white-painted tubular steel trusses. Flexible hinges and silicone-based glue ensured that the building could withstand an 8.0-magnitude earthquake. (Schuller, a disciple of Norman Vincent Peale, raised a lot of the money for construction by inviting donations for individual panes of glass.) The result upon completion in 1980 was the largest glass building in the world.

I loved the idea of a natural ventilation system. We began by recommending the use of silver-coated reflective glass that would protect against the heat of southern California. Only ten percent of the sunlight shines though, and yet even on a hazy day, the reflected light keeps the cathedral's interior brightly lit.

I decided to keep the place cool with natural ventilation. For that to work, we needed operable windows on all sides, and the air had to be brought in low down in the building and to exit high up. The architectural team louvered the windows for easy movement. We came up with a computerized program that reacts to solar and wind conditions and, as required, opens coordinated banks of windows for cross-ventilation. The breeze then

The Crystal Cathedral, Garden Grove, California (Philip Johnson and John Burgee, 1980)

drifts through the building, rising as it warms. When a certain temperature is reached, upper windows thermostatically open to exhaust the warm air, drawing in fresh supplies through natural convection.

Depending on the weather, two 90-foot-high doors can be swung open to the surrounding landscape so that it seems as if services are being conducted outdoors. We used no air-conditioning except at Dr. Schuller's stage. Because his sermons were televised, there were so many lights shining on him that he was baking in the heat. We put in localized air conditioning under the stage to make Dr. Schuller more comfortable.

Thanks to its various mechanical systems, the cathedral, despite its large size and glass walls, consumes very little fuel. It's not a perfectly controlled environment, but Nature did a good job of maintaining comfortable conditions. Ironically, the building's efficiency has outlived the ministry it was designed to serve. After Dr. Schuller retired in 2006, his church began a downward slide that ended in bankruptcy court several years later.

The Crystal Cathedral was sold in late 2011 to the Catholic Church. It remains to be seen how this Modernist showcase, fresh breezes blowing through its open walls, will accommodate traditional liturgical requirements.

Cosentini was quick to recognize the value of new computer technologies for recycling energy, sometimes without using any computer programming at all.

We were able to maximize the possibilities at PPG Place, an ambitious 6-building project, kind of like Pittsburgh's Rockefeller Center, which was built in a depressed part of the city to spur transformation from its sooty industrial past. Like the many other Postmodern buildings that Philip produced in the 1980s, this Gothic design draws inspiration from well-traveled historical models. Its iconic 40-story tower bears a strong family resemblance to Victoria Tower, Big Ben's stouter and lesser-known sister at the Houses of Parliament in London.

Pittsburgh Plate Glass was the client, and it wanted the project to showcase the company's new generation of high-performance curtain walls. We were able to devise a system for year-round high-energy efficiency despite the fact that the exterior was all glass—19,750 pieces to be exact—totaling more than a million square feet. Silver insulating reflective glass reduces heat gain in summer by reflecting sunlight away from the surface, while it reduces heat loss in Pittsburgh's cold winters, using a thermal barrier to isolate the interior and exterior.

We went beyond this to capture and recycle the tremendous heat gain generated by PPG's massive computer center. In cold weather, we recycled through the same water storage system that was used for air-conditioning in the summer. Thus, we circulated the cooled water in the summer and the heated water in the winter. The scheme worked so well that it wasn't until the temperature fell below three degrees Fahrenheit that we ever had to use auxiliary heat pumps. There was, however, a

PPG Place, Pittsburgh, Pennsylvania
(Philip Johnson and John Burgee, 1984)

glitch. Several years after the building was completed, PPG moved its computers to another location. Our source of heat was gone and we had therefore had to install boilers to heat the place.

It is not only big buildings that present interesting energy challenges. In the early 1970s I worked on an early solar-energy residence with Dr. Mária Telkes, of the University of Delaware's Institute of Energy Conversion. Dr. Telkes, who was known as the "Sun Queen" for her contributions to solar energy research, had designed the Dover Sun House, in Dover, Massachusetts, in 1948 while a professor in MIT's Department of Metallurgy. This pioneering "house of the day after tomorrow" used Glauber's Salt, or hydrate of sodium sulfate ("miracle salt"), stored in big metal containers on the roof. During sunny days, the salt would get very hot. When the sun went down, the salt would liquefy and, as it melted, release its stored heat, which was then circulated by fans through the 5-room house. The Dover House system failed after a few years but in time the problems were resolved.

Years later, I worked with Dr. Telkes on Solar One, the first house to harvest solar energy in a total-system approach to generate both heat and electricity for domestic use. This experimental house, built by the University of Delaware at Newark in 1973, still stands on South Chapel Street, and its integrated solar harvesting system still works. Of course, solar technologies have developed a great deal since then. We now use high-performance photovoltaic panels. When photons in sunlight hit the chemicals in those rooftop panels, electrons are activated to a higher state of activity, and an electrical charge is created. That electricity can be stored for future use. It is a clean and elegant way to create electricity.

With the old cells, we'd get about one watt per square foot. With today's photovoltaic panels, optimally installed, we get 11 watts per square foot, and there are even some panels that can produce 16 or 17 watts per square foot, although they're quite a bit more expensive. In every new project we do we try to use PVs or other energy-efficiency devices in at least part of the building.

Solar One, Newark, Delaware (Mária Telkes, 1973)

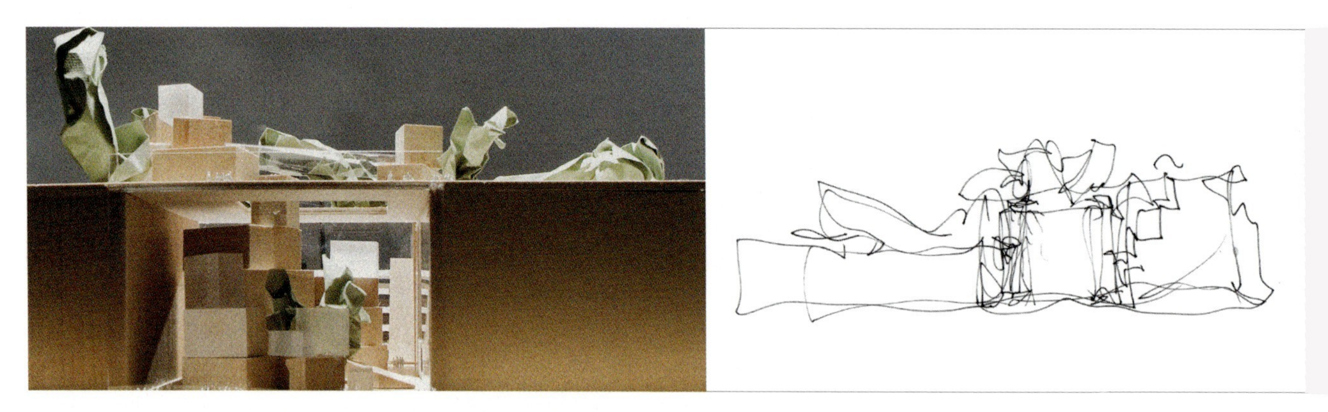

Left to right: New World Symphony model (Frank O. Gehry, 2011), Guggenheim Bilbao sketch (Frank O. Gehry, 1993); right: Gehry (2011)

FRANK GEHRY

"Marvin is one of the most loyal consultants in our profession. He has always been a great collaborator and an enthusiastic supporter of exploring new ideas."

Frank Gehry

I have had a lot of fun working with Frank Gehry over the years. We were the mechanical engineers for many of his buildings, including the Guggenheim Museum Bilbao (1997), the Richard B. Fisher Center for the Performing Arts at Bard College (2003), Walt Disney Concert Hall in Los Angeles (2003), IAC Headquarters in New York (2007), the Lewis Science Library at Princeton (2008), and the New World Center in Miami, which opened in 2011.

In 2001 Princeton University received a $60 million gift for a new science library from an alumnus Peter B. Lewis, who, as chairman of the board at the Guggenheim Foundation in New York, had worked with Gehry on Bilbao. Shortly after the job came in, I went to see Frank in California and asked about the building's design. He pointed to a shelf where there were models of various campus buildings and another model that seemed to be protected by corrugated paper. I tried to remove the covering so I could get a better look but Frank called out, "No, no. That's what the building looks like. Don't touch it."

As it turned out, this early scheme was too expensive to construct, so Princeton and Gehry had to compromise on a scaled-down design. The basic "think outside the box" concept remained.

Our work at Princeton had two components. The first was to update the various systems of the existing library: air-conditioning, lighting, communications, and so on. Those then had to be integrated with the systems of the new building, since sections would be joined into a large main gallery. It was a big and complicated job, yet there's little evidence of our work other than increased efficiency, comfort, and convenience.

Recently, we worked with Gehry on the New World Symphony in Miami, which opened in January 2011. It is totally different from the Walt Disney Symphony Hall in Los Angeles, which Frank completed in 2003 (16 years after the project began). In fact, the New World Symphony is unlike any concert hall I'd ever seen before. Gehry designed it for his old friend Michael Tilson Thomas, the great composer and conductor, who wanted a flexible space for more experimental musical performances.

Frank delivered a 756-seat auditorium fully equipped with fourteen stage configurations, and retractable seats, together with four performance platforms, like little satellites, so that music wouldn't be restricted to just the stage. The interior is wonderfully animated by suspended sail-like acoustical panels that double as video projection screens and also, uniquely, by ever-changing natural light.

I had my doubts about a major design element even before construction started. The plans called for a huge window behind the stage, which I had never seen before, since music halls are usually located deep inside buildings to protect them from sound, vibration, and all the other distractions of the outside world. I doubted that external noise could be blocked, but a specially engineered multilayer glass laminate made it possible. The window, and the floods of daylight it admits, change the whole character of the performance space.

In designing a concert hall, the crucial factor is not the mechanical or structural engineering or even the architecture, but the acoustics.

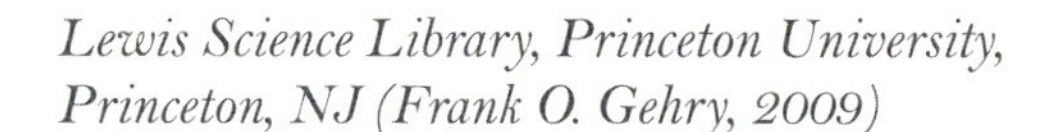

Lewis Science Library, Princeton University, Princeton, NJ (Frank O. Gehry, 2009)

New World Symphony, Miami, FL (Frank O. Gehry, 2011)

The acoustician often has as much influence as, if not more than, the architect over what the music chamber ends up looking like. He is concerned that, wherever you are seated, you can hear high frequencies and low frequencies at the same time. For example, an architect may design curved walls for a concert hall but the acoustician might veto the plan because of potential sound distortion. Architect or acoustician, who's in charge?

This, of course, explains why concert halls are so very difficult to build and why so few truly great concert halls have been created. According to audiences and musicians, and also in the humble opinion of this mechanical engineer, Gehry succeeded in Miami.

Sound engineers work with computers today but they used to rely on their own inner ear and sound memory. I remember working on the Krannert Center for the Performing Arts at the University of Illinois at Urbana-Champaign. Max Abramovitz was the architect and Cyril Harris, one of the best acousticians of our time, was in charge of the sound.

Cyril and I flew out together from New York. He was carrying a suitcase and when we got to the music hall, he opened it up and took out a small chromium-plated cannon. He set it on the stage, loaded it with ten-gauge blank shells, and fired. I couldn't believe what I was seeing. Cyril stood very still for a moment carefully listening to the reverberations and then smiled and said, "Perfect." He would never get past airport security now.

One of the world's most troubled concert halls has got to be Avery Fisher at Lincoln Center in New York. Max Abramovitz originally designed it as a traditional 2,400-seat "shoebox," the rectangular form preferred for optimum sound. But bowing to public pressure, the scheme was enlarged at the last minute to include several hundred additional seats and construction proceeded full speed ahead—without having made all the necessary acoustical refinements.

Philharmonic Hall, as it was originally named, opened in 1962. No one liked the sound: not the conductor, orchestra, audience, or critics. Numerous adjustments were made over the years but the hall's problematic acoustics continued.

Not so Carnegie Hall, which, as a rival of the Philharmonic's new home, had been slated for demolition but which, given the turn of events, gained a new lease on life.

In a funny kind of sequel, my wife, Ruth, and I attended a concert at the Mann Auditorium in Tel Aviv some years later and the sound was absolutely glorious. When I asked who had designed the acoustical system, it turned out to be BBN, the same firm responsible for what we now know as Avery Fisher Hall at Lincoln Center. I've come to appreciate over the years that, job to job, the people, place, program, technologies, politics, and budget are all different, and that things are never so simple as might initially appear. Every building has its own unique story.

Because of the significant challenges, many architects hold their breath before taking on a concert hall, and gasp at the very thought of ever doing another. Gehry is more intrepid about risk-taking performance spaces, as he is a lifelong devotee of music, "a Paganini among architects."

Richard B. Fisher Center for the Performing Arts at Bard College, Annandale-on-Hudson, New York (Frank O. Gehry, 2003)

I worked with Frank on the Richard B. Fisher Center for the Performing Arts at Bard College in New York's picturesque Hudson Valley. Because the building is more circular than rectangular, Yasuhisa Toyota, the acoustics engineer, wanted a chamber at each side of the auditorium. The sound was meant to go into the chambers and reverberate out into the main hall the way it would from a cathedral apse. Of course, that meant that the ventilation system we designed also needed to accommodate the unique design.

When you drive onto the Bard campus you see this beautiful little stainless steel structure kind of fluttering over the hillside. The surprise comes when you go inside and find a symmetrical proscenium concert hall, together with a smaller, more flexible state-of-the-art theater, well-used rehearsal studios, and lots of other square-edged support spaces. Another surprise comes at the back of the building, where the exposed concrete walls and loading dock are the unadorned results of an exhausted budget.

Gehry has been criticized because his buildings don't fit in with their surroundings. He famously said "I don't do context." In most instances, that has been all to the good. Gehry is one of the great place-makers of our time, and his buildings create their own context. The Walt Disney

Garden Entrance: Walt Disney Concert Hall, Los Angeles (Frank O. Gehry, 2003)

Concert Hall in Los Angeles, which I also worked on with Frank, is a prime example of how a single magnificent building can completely change the area in which it was built.

Even more profound was what Gehry did for Bilbao, Spain, an old Basque city that became a leading 19th-century industrial center before being forced to its knees by economic crisis in the 1980s. Bilbao's ship-building industry moved to Korea, unemployment soared, and violent terrorism marked the day. The city's river was polluted, sewage was dumped on its shore, and historic buildings and streets crumbled.

In 1991, in association with Thomas Krens, director of the Guggenheim Museum in New York, the city of Bilbao commissioned Frank Gehry to design what Philip Johnson later called "the most important building of our time." Overnight, Gehry's museum rescued Bilbao from the grime of its industrial past and transformed it into a mecca for culture and tourism. The museum not only forever changed the city, it redefined the role of public buildings and how we see them. Now, when I visit Bilbao I see Frank's building billowing out at the end of the street and one of its sculptures smiling at me. I am always moved by the power of architecture to bring about social change.

"Architecture is a small piece of the human equation," Gehry explained, "but for those of us who practice it, we believe in its potential to make a difference, to enlighten and to enrich the human experience, to penetrate the barriers of misunderstanding and provide a beautiful context for life's drama." Frank himself was swept along by the potential. Though he had used metal cladding before, this was the first time he used titanium on the outside of a building. At the museum's dedication in 1997, the local contractor presented Frank with a jacket made of titanium. I'm sure he still cherishes it, although I'm not so sure how much he wears it.

It's important to understand that Gehry works on his buildings from the inside out. He designs what the user wants and then surrounds them with his pyrotechnics. For example, at Guggenheim Bilbao, Frank designed straight-walled, rectangular galleries like a series of stacked boxes and then covered them with the fantastic glistening shapes we now see.

What made all this possible was a type of computer software called CATIA (the French acronym for Computer Interactive Aided Three-dimensional Application), which had originally been used in designing the continuous curvature of aircraft. Gehry employed the program at Bilbao for the first time. Its advanced modeling capabilities created new opportunities for Gehry and also for other expressionistic architects through the spinoff company, Gehry Technologies, that he established to share the technology.

Eventually, Bilbao enjoyed the work of other great architects, including César Pelli, Rafael Moneo, and Arata Isozaki. Norman Foster, in an early commission in 1988, designed a

Guggenheim Museum Bilbao, Bilbao, Spain (Frank O. Gehry, 1997)

fully integrated subway system with distinctive glass entry pavilions known today as Fosteritos. And Santiago Calatrava designed a splendid footbridge over the river Nervión as part of a much broader campaign to reclaim the city's riverbanks with beautiful parks and play areas. Bilbao is now a very magical place whose citizens are immensely, and justly, proud of their accomplishment.

Even when we were doing the museum, I sensed that the workers were so happy to be making a living and to have something interesting to build that they went out of their way to accomplish whatever we designed. And although at some point the Basque separatists strongly objected, the Bilbao community really supported us. The Buildings Department, for example, had no regulations for constructing a place of public assembly like this one. So instead of being constrained, we were able to take advantage of everything we'd learned as engineers and architects in Europe and America and put it to good use here. The people of Bilbao accepted what we were doing and got a great facility, and worldwide acclaim, in return.

Several years later Gehry was giving a lecture at Low Library at Columbia University and he had invited many of his colleagues to the event. Max Abramovitz was on Columbia's Board of Trustees and I was sitting next to him. Just in front of us was the Argentine architect Rafael Viñoly, as well as other distinguished people in our field. Gehry was on the stage talking about his work with photographs projected on the screen beside him. The trouble was that from where he was standing, Frank could not really see the screen, so he looked at the audience and said into the air, "Hey, Marvin, what job was that?" It was unmistakably Guggenheim Bilbao.

One of the most successful recent office buildings in Manhattan is the headquarters of IAC, which Gehry built in Chelsea in 2007. This

freestanding building offers a welcome change from all the typical four-square blocks erected in the city. Gehry wanted the windows to evoke great wind-filled sails overlooking the Hudson River, so he enclosed the 10-story building in swooping floor-to-ceiling glass walls. All well and good, but we had to do something to make the building usable because, situated on the far west side of the city, facing the setting sun and just above the river's glare, the solar load would have been extreme.

Rather than using sills or some other external shading apparatus, we embedded in the glass (all 1,437 panes of it) tiny ceramic frits that cut out 50 to 75 percent of the sun's heat and light. During the day, the milky white windows look like bright pleated sheets of metal. At night, light shines through the frit and the building becomes transparent. It's just terrific. During office hours, workers enjoy sprawling views through clear vision bands but, because of the frit, the space does not heat up. We were able to cut down on the air-conditioning load and, at the same time, created a great nighttime light show, like some big fluttery lantern along the water's edge.

When IAC opened in 2007, I asked a receptionist how she liked it. She answered, "It's the best place I've ever worked. I really enjoy coming here every morning."

That was a real compliment. I might have added that the pleasure was mine, as I really enjoyed working with Frank, and have always felt privileged to help to bring his wonderfully creative buildings to life.

IAC Headquarters, night and day, New York (Frank O. Gehry, 2007)

THE MIDDLE YEARS: ARCHITECTS GALORE

A few of my friends…

Do you know them?

Row 1: Max Abramovitz, Edward Larrabee Barnes, Gordon Bunshaft, John Burgee, Santiago Calatrava, Peter Chermayeff, David Childs, Peter Claman

Row 2: Henry N. Cobb, Araldo Cossutta, Joseph Fleischer, Lord Norman Foster, Bruce Fowle, Robert Fox, James Ingo Freed, Frank O. Gehry

Row 3: Arthur Gensler, Peter Gorman, Michael Graves, Charles Gwathmey, Hugh Hardy, Walter Hunt, Helmut Jahn, Philip Johnson

Row 4: Gene Kohn, Morris Lapidus, Jill Lerner, Ralph Mancini, Richard Meier, William Pedersen, I. M. Pei, César Pelli

Row 5: James Stewart Polshek, Alan Ritchie, Kevin Roche, Nancy Ruddy, Peter Samton, Robert A.M. Stern, Marilyn Jordan Taylor

I was very fortunate to begin my career working with some of the best architects in the country. I am happy to say that my good fortune continued as I had the opportunity to work with even more great architects on even more great buildings. Thinking back, it is appropriate to start with Gordon Bunshaft of Skidmore Owings & Merrill (SOM), since I started out with him as project manager for Lever House, the pioneering glass and steel skyscraper that became an instant icon of the International Style of architecture.

Gordon went on to design some of the city's most elegant office buildings. My favorite, which I also worked on, was 9 West 57th Street, where the glass north and south facades slope up from the sidewalk. With this sculptural building, right in the middle of the block, Bunshaft really took architecture outside the box, and totally transformed the closed, heavy masonry street wall into something much more open and lively. Twice: He used the same sloped form fifteen blocks south at the W. R. Grace Building on 42nd Street.

Gordon simply recycled his rejected preliminary scheme for 9 West and used it for its sister building. The two 50-story towers were constructed simultaneously and completed in 1974. Sheldon Solow, the owner of 9 West, was unhappy that the other building duplicated the unique shape of his new landmark. The truth is that a decade earlier, in 1963, Harry Cobb had come up with a similar sloped solution for Finance Place, an unexecuted building for the stock exchange, in fact, on the site where SOM built the U.S. Steel Building, also in 1974.

Gordon later explained that his sloped design resulted from his attempts to comply with the complex New York regulations for zoning, setback, and sky exposure. For me, the contour was akin to fitting a square peg in a round hole, since the facade was curved and the induction units next to the windows were straight.

SOM had a lot of very good architects who did some wonderful buildings, but everyone wanted Gordon Bunshaft; for decades hedefined SOM and sleek corporate modernism. After Gordon retired, Skidmore felt the need to recruit and brought up David Childs and Marilyn Jordan Taylor from Washington, D.C. They reinvigorated the firm. David and Marilyn are both great architects and it's been my pleasure to work with them on many of their buildings.

Another very influential architect was William Tabler, the number one hotel designer in the United States, although he is largely unknown outside the profession. Tabler was responsible for more than 400 hotels and, as business

9 West 57th Street, New York (Gordon Bunshaft/SOM, 1974)

travel increased in the years following World War II, he created a new corporate image of streamlined efficiency, especially for the Statler and Hilton hotels.

Tabler played a major role in developing the Hilton Inn chain, for which he developed an economical modular building system. Everything was in 4-foot multiples so full sheets of drywall could be installed, one right next to another, without ever having to cut them. He also developed the Tablin sink system, which combined sink, countertop, and toilet into a modular whole and became a standard for Hilton Inn across the country.

The biggest job we ever did with Tabler was the New York Hilton on Sixth Avenue, near Rockefeller Center, which opened in 1963. With 2,153 rooms, it was, and still is, the largest hotel in New York. The original architect was actually Morris Lapidus, with whom we also worked, but when it was discovered that Lapidus was simultaneously involved with the competing Americana Hotel nearby, Tabler was brought in to take over the Hilton. He designed a 5-story masonry base topped by a 46-story blue glass and steel tower. The faceted facade allowed all the mechanical systems to run outside the structure, creating larger guest rooms with the additional space.

The hotel client was Rock-Hil-Uris, a three-way joint venture of Laurence Rockefeller, Conrad Hilton, and the Uris brothers. I will never forget the building's official beginning when Harold Uris broke more than ground. He picked up the stone sample that Bill Tabler had selected for the Sixth Avenue facade and dropped it, shattering it into pieces on the ground. "Too expensive," he said, "Use a different stone." Just like that. I'd never seen anything like it before, or since.

Tabler was the associate architect on the New York Hilton hotel along with Harrison & Abramovitz, another firm we worked with on a number of big projects. Most of my day-to-day contact was with Max Abramovitz, a small, very quiet and unassuming guy, born and bred in New York. Harrison & Abramovitz did a lot of big, high-profile jobs thanks to Wallace Harrison who just so happened to have close connections with the Rockefellers.

Curtain wall, New York Hilton (William Tabler, 1963)

In the same way that Tabler was the country's leading hotel architect, Emery Roth was the leading residential architect in New York, later branching out into curtain wall office buildings. He and his two sons, Julian and Richard, did scores of buildings, many of them with us. Emery Roth & Sons was a real institution, known for consistently producing highly marketable buildings with maximum rentable area and windows that didn't leak. When the designers Larry and Harry Harman left to open their own firm, we worked with them, too. In fact, I worked on their first big job, a Fifth Avenue apartment house, at 75th Street. Years later I moved into this building and asked for a set of plans. I was surprised to see that I was the draftsman! I guess I couldn't complain about anything. Life is filled with funny surprises, and after you've been in business for a half century or so, things tend to overlap and connect.

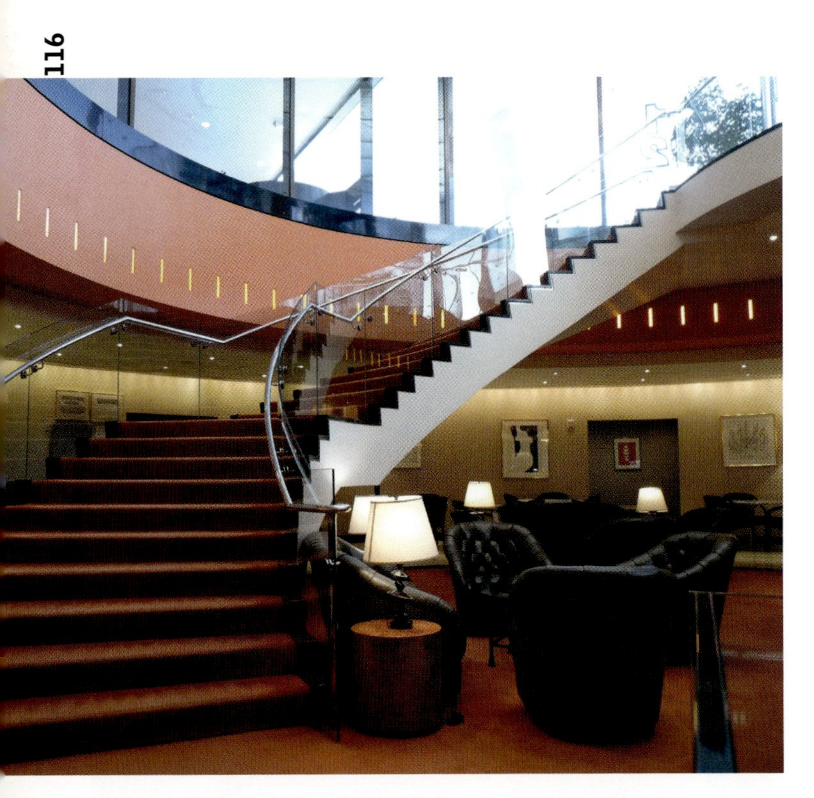

Through the years I always made a point of attending meetings with architects. I made it my business to understand their intentions, to know what they wanted, how they expected their buildings to perform, to understand client requirements. I didn't wait for the architects to just give me their drawings so I could superimpose my systems. I anticipated needs and shared my ideas and suggestions. In the process we became friends.

One of my very good friends is Hugh Hardy. I met him while working with Hardy Holzman Pfeiffer but I can't pinpoint on which job, maybe the Grand Opera House in Texas (1974) or Boettcher Concert Hall in Denver (1978). It doesn't really matter because it seems like we've always known each other. Hugh and his wife, Tiziana, also an architect, are just the nicest people in the world. He's a very talented guy and does things with effervescence, all of which adds a special quality to his projects. He does lots of performing arts facilities (Hugh himself is a talented pianist) and also high-profile restoration projects like the beautiful Central Synagogue on Lexington Avenue at 55th Street, after it was destroyed by fire.

Given the prominence of his other projects, I was surprised that Hugh Hardy agreed to design 8½, a restaurant in the basement of

Brasserie 8½ at 9 West 57th Street, New York (Hugh Hardy/H3 Hardy Collaboration, 2003)

Gordon Bunshaft's sloped tower at 9 West 57th Street. Hugh brought this empty space to life with the same high quality and drama that characterize his other projects. It's one of my favorite restaurants in all of New York.

Shortly after 8½ opened, Hugh invited Ruth and me to dinner. I remember sitting there as Tiziana came down the great curved staircase that makes such a grand entrance down from street level. Hugh called out, "Champagne, Love! Champagne for everyone!" And then he turned to me and said, "Marvin, you are among the first to know that my partners and I are separating and that I am starting a new firm (H3 Hardy Collaboration Architecture). Champagne for all!"

It isn't often that one witnesses a firm's beginnings, but that pretty much happened to me once before in 1978 when I helped Fox & Fowle get established. I provided space in our office for a couple of years until a steady flow of work came their way and they moved out on their own. To this day, when I see Bruce Fowle at some public function or when he comes to my annual party, he looks my way and says, "Marvin put me in business." I was happy to help. That's what friends do.

More than any other architects, Fox & Fowle played a defining role in the recent transformation of seedy, crime-riddled Times Square into a showcase of green corporate architecture with buildings like Condé Nast at 4 Times Square (1999) and the adjacent Reuters Building at 3 Times Square (2001). In time the two architects separated and Bob Fox formed

11 Times Square, New York (FXFowle, 2010)

a new firm with Richard Cook. (Cook once claimed that I taught him everything he knew about buildings! That's very high praise from such a talented designer.)

My good friend Bruce Fowle, with his new firm FXFowle, continued to design sustainable buildings in the Times Square area. He joined with Renzo Piano on the New York

The Portland Municipal Services Building, Portland, Oregon (Michael Graves, 1982)

Times building (2007) and then designed 11 Times Square right next door. We worked with FXFowle on the latter 40-story glass tower, which is an absolute model of energy efficiency. This very, very green building had everything going for it, except timing.

The groundbreaking for 11 Times Square took place in 2007 when Manhattan's commercial real estate was sizzling but by the time the building opened in 2010 the economy had reversed and this empty spec building became a symbol of the troubled real estate market. At one point, the search for new tenants had us working with Peter Chermayeff on the design of a 7-story aquarium for the bottom floors of the building. But in the end, financing fell through.

But I'm getting ahead of myself. The early 1980s was the heyday of Postmodernism, and we were involved in the movement's first major icon, the Portland Municipal Services Building, which opened in 1982, a couple of years before Philip Johnson's famous Chippendale-topped AT&T Building in New York. It was Michael Graves's first big commission, and it put him in the center of heated national debate. There didn't seem to be any gray area in this "jolt of color." People either loved it or the opposite. I thought the 15-story building was great.

There simply was nothing of this quality in the city, and it became the touchstone of many new downtown developments. The Portland Building was so important architecturally and as a cultural icon that it was added to the U.S. National Register of Historic Places, before it was fifty years old.

We've done a lot of work with Michael Graves since Portland. I just love him, and have always found him to be a great educator. He is so knowledgeable that everything he says is worth learning. Graves doesn't talk just about architecture, but about culture and the big picture. He's so filled with life, even after being confined to a wheelchair in 2003; he still works as hard as ever.

I remember seeing him just last year with Peter Eisenman at Cooper Union. Both architects were lecturing about how things have changed over the past decades. Eisenman was giving it pretty good to the young architecture students, insisting that when he was their age, he'd have a pencil in his hand and would automatically think in three dimensions. "Today, you just look at a flat screen," he said. "I don't know how you could possibly design a building without seeing in three dimensions." Graves agreed (so did I) but he approached the subject from a different perspective. Where Graves has been having an impact lately is on accessibility and universal design. He is quick to point out the positive, and to explain that he probably would never have focused on health care design if he hadn't gotten sick. Graves is such a wonderful man! I really admire him.

The lecture at Cooper Union got me thinking about the old days at Cosentini. We used to spread out our drawings on a big light table, and then layer the structural and architectural drawings on top to see where the conflicts were. These days, it's all done by computer. Some people in my office don't even know what a light table is, and have never used or even seen a slide rule! It's unbelievable how things have changed.

Another architect I admire, is César Pelli, who designed the residential tower for the Museum of Modern Art in 1984. It was part of MoMA's fiftieth anniversary expansion program and was designed to help pay for the museum's expenses. A lot of people were upset about the 52-story mid-block tower but it never seemed to rattle César. He is such a gentle man, so quiet and unassuming. I started working with César when he first left Eero Saarinen's office, a half century ago, and we've done a lot of buildings together ever since. Now we frequently work with Rafael Pelli, who is a fine man, just like his father. In all the years I've known César I never heard him say a bad word about anyone. He always talks about the good and the positive.

Like Philip Johnson, who had an apartment in MoMA's tower, my involvement with the museum continued for decades. In 1997, at the opening of Guggenheim Bilbao, I ran into James Snyder, MoMA's longtime deputy

MoMA Tower, New York (César Pelli, 1984)

director, who had been responsible for Pelli's expansion. It turned out that Snyder had left MoMA the preceding year to become director of the Israel Museum in Jerusalem.

"The one with the Dead Sea Scrolls?" I asked. "Yes," said Snyder with obvious surprise. "How did you know?" Well, it just so happens that back in 1965, I worked with Bartos & Keisler, the original architects of the Israel Museum. We were the engineers for the building and for the Shrine of the Book. Some months later, Ruth and I visited Jerusalem, and James Snyder showed us around the museum. And there were the 3,000-year-old scrolls, still inside the climate-controlled tube that I had designed for them.

Israel Museum, Jerusalem (Bartos & Keisler, 1965)
Dead Sea Scrolls, Shrine of the Book

I could go on forever about the architects I've worked with. There are so many that their names could, well, fill a book. People like Ed Barnes, Norman Fletcher, Charles Luckman, Victor Lundy, Charlie Gwathmey, and Bob Siegel, and all the wonderful residential architects who came after Emery Roth and formed a new generation of residential designers: Philip Birnbaum, Peter Claman, Costas Kondylis, Alan Goldstein, Ismail Leyva.The list goes on. A lot of my friends started out straight after school and later formed prominent new firms, like Moshe Safdie, with whom I worked on his first big project, Habitat 67 in Montreal.

I've lived with a lot of architects through their evolving identities, like Lew Davis who founded Davis Brody Juster & Wisniewski, then Davis Brody, followed by Davis Brody Bond. Now that Lew, Sam Brody, and Max Bond have all passed away, Steve Davis runs the Davis Brody Bond Aedas firm. Other architect friends who have transformed over the years include Gruzen Samton Steinglass, Swanke Hayden Connell, Cooper Robertson + Partners, plus there are other classic firms like Eggers & Higgins, Kahn & Jacobs, and Mitchell Giurgola that have grown with their original names intact.

There are plenty of architects who have worked with my partners or other engineers in my firm, who nonetheless have become my personal friends, like Richard Meier, who used to live near me so I'd see him on the street and Bob Stern who I like very much. Basically I know Bob through Yale. As dean of the School of Architecture, he asked me to teach there, which I did for several years until I opted for the closer commute to Cooper Union down the block! The point is that it's personal, not just business.

As a rule, I'd say that I feel closer to architects than to engineers, who too often live up to their reputation for being rigid, uncompromising, and demanding. Architects, by contrast, are warm and outgoing and most often have a good sense of humor. I guess it also helps that they're not my competitors!

One of the great exceptions to this rule is Ed Messina, a structural engineer who heads up Severud Associates in New York. I had a lot of fun with him while working on the National Museum of the American Indian, principally designed by Douglas Cardinal, a Blackfoot Indian, and Johnpaul Jones, a Cherokee/Choctaw. Once a month we would go to Washington for a meeting with tribal leaders who were invited from across the country to share their vision of what the museum should be and how it should look. The building's earthy architecture, unlike anything else on the National Mall, took shape as a direct result of these sessions.

At the end of each meeting the elders would pass around a peace pipe and everyone would take a puff. Keeping the peace was no simple matter in a building that took fifteen years to complete. The best part of this long schedule is that I had lots of time to get to know Ed Messina better.

When the building finally opened, in 2004, Ed came over to me, gave me a little barb, and said, "The air-conditioning in the building isn't working. It's too hot in here." I replied, "The floors are bouncy. The museum needs a good engineer." I have often found that a little humor goes a long way in helping even the most difficult situations. In this case, Ed and I were just having fun. The museum's floors are as firm as could be, and the building systems work so well that people don't even think about them.

National Museum of the American Indian, Washington, D.C. (Douglas Cardinal and Johnpaul Jones et al., 2004)

Over the years I have gone to many interviews with architects and engineers. I remember going to so many with Irwin Cantor of Cantor Seinuk structural engineers that we would trade off and he would discuss the mechanical systems and I would discuss structure. We knew each other so well, and it kept the whole process interesting and fresh.

One of the especially gratifying parts of growing up with architects is seeing young men and women blossom and come into their own. I think immediately of people like Bill Pedersen, who began as a young designer out of I. M. Pei's office, and Gene Kohn, Bill's partner at KPF. These young architects designed wonderful buildings. When things slowed down in the 1980s, Jill Lerner joined KPF and brought in public work, like schools and hospitals, and Anthony Mosellie brought in important airport projects. KPF expanded globally and is now one of the most prominent firms in the world.

We have done lots of work over the years with KPF, including a recent 17-story vertical campus at Baruch College in Manhattan near Gramercy Park. The incredibly complex mixed-use building covers an entire city block and includes classrooms, offices, labs, a 500-seat auditorium, an athletic center and swimming pool, a theater, conference rooms, dining, and student services, all around a series of central atriums. Basically, it pulls together everything needed by the business and liberal arts schools in one big high-tech building.

Even more complex is the new federal courthouse that KPF did in Foley Square. What an incredible building this is! It's a building with dozens of courtrooms, judges' chambers, ceremonial spaces, and support for the district attorney and U.S. Marshall. It also includes more than 50 detention cells, all in a 27-story office tower with state-of-the-art communications, high-tech security, and life

Newman Vertical Campus, Baruch College, New York (Kohn Pedersen Fox, 2002)

safety systems, everything very flexible and energy efficient.

The challenge was heightened because the building had to be operational throughout the day as you never know when things are going to happen. The big thing for us was temperature control, with all the round-the-clock comings and goings, and certain occupants who kept trying to open the windows, which, of course, changed the building climate. For me, all deliberations came down to air-conditioning!

As the economy has shifted and architects have been drawn east, we've also been doing a lot of work in Asia and the Middle East. In Shanghai, right next to KPF's World Financial Center, is SOM's Jin Mao Tower, and now under construction is the adjacent Shanghai Center, a 2,074-foot-high tower that we're doing with Gensler. Upon completion in 2014, this vertical city will be the tallest building in China and the second-tallest building in the world. Things certainly have changed!

I remember Art Gensler and his wife back in the 1960s and '70s. They were doing interior design, not architecture. Then slowly things started to shift and they began bringing in some commercial buildings. Their office has never looked back, and now, through the efforts of such architects as Walter Hunt, Gensler has achieved global prominence. We are also working with them at the other end of the height spectrum on a new low-rise corporate campus for the Novartis Pharmaceutical Corporation in New Jersey.

Shanghai Center, Shanghai (Gensler, in progress)

Actually, Walter Hunt just recently retired and was honored by the AIA at the Center for Architecture in New York. In fact, I gave a speech in his honor, although I have to admit that the way this happened came as a bit of a surprise. I had just gotten off the plane after visiting Panama with Ruth. We got back to our apartment, and waiting for me was an invitation to Walter's event. It was bright red, so I couldn't miss it. And there I was, listed as one of the speakers, scheduled to begin barely an hour later! Off I went with not a moment to spare.

I've been fortunate in my career to have worked on some of the biggest and most important buildings of our time. The drive toward big, tall, and high-profile buildings continues to this day, not just in developing countries but around the world. The problem is that the extremes often don't make their way to completion, usually, but not always, due to financing.

That was certainly the case with the Secaucus Junction, which we did with my personal friend Peter Gorman, partner of Brennan Beer Gorman. The ambitious plan called for a multilevel regional transit hub to link ten of New Jersey Transit's eleven rail lines. The goal was to improve access within the state and to expedite the commute into New York.

The project was designed for 3.5 million square feet of commercial space, including five new high-rise hotels and office towers

Frank R. Lautenberg Secaucus Junction Station
Secaucus, New Jersey
(Brennan Beer Gorman, 2003)

that would overlook and complement Manhattan's skyline. The transportation center was built first (2003), but then nothing else went ahead. There is now some discussion of extending New York's #7 line out to the hub, which would be the first time the subway ever went beyond city borders. This would bring in lots of people to Giants Stadium and the Meadowlands. For the time being, it remains pretty lonely in Secaucus.

Another important project that didn't go ahead was the Atlanta Symphony Center, which we were working on with Santiago Calatrava. It was a terrific scheme with great steel lattice wings that opened and closed over the soaring upper lobby. The dramatic scheme, which used all of Calatrava's skills as architect, engineer, and sculptor, was released with great fanfare back in 2005, but four years later the $300 million project was scrapped. Early on in the process we met with Calatrava in our conference room and he sketched his concept for us. I asked him to sign the sketch, had it framed, and now it hangs in my office, a colorful reminder of a building never realized.

Another Calatrava project from right around the same time was a bridge for the real estate developer George Klein. He had gotten the rights to build some apartment buildings in Brooklyn, but since there was no easy access to the site, he hired Calatrava to design a new crossing from Manhattan. So far nothing has come of the apartments or the bridge, but every so often we hear rumors of reactivation.

Atlanta Symphony Orchestra
(Santiago Calatrava, 2005; unbuilt)

If anything does come of this project it will be my son, Doug, who will take it forward.

I have had a very full life, filled with great friends and experiences, and now the future of our industry moves ahead in collaboration with a new generation of architects, including talents like Enrique Norten, Rem Koolhaus, Chip Calcagni, John Cetra, Jon Picard and Bill Chilton, among many others. We look forward to working with them and with other architects like Gary Handel, Jim Davidson, and Luigi Russo, and increasingly with a new group of talented women architects who are making a real name for themselves in the profession, people like Marilyn Taylor, Nancy Ruddy, Jill Lerner, and Mary Jean Eastman, among others. Exciting times are ahead.

RECENT YEARS AND GOING FORWARD

DOUGLAS C. MASS, P.E.

I joined Cosentini in 1983 at the beginning of a building boom. Things were very different than they were when my father started out in the early 1950s, working closely with all the architects he has talked about in this book. In those days, architects were great builders, particularly since there had been very little construction for decades. Architects didn't necessarily understand how to fit building systems into their designs, and because air-conditioning was in its earliest commercial stages, they would work with the mechanical engineers and think through problems together. There was a lot of creativity, and certainly in my father's case, a lot of important innovation. The mid-century A&E process was all about integration and collaboration.

By the time I arrived at Cosentini, the nature of the building design process had changed. Architects were very busy and in effect said, "Here are my plans; I'm done. I haven't really given a lot of thought to your building systems. Put them in and make them work within my volume... and don't get in the way of my design." Numerous projects came into the office, and we'd produce them one after the other.

It was a time when style was king. Prominent architects and their clients wanted to expose this, sculpt that, and cantilever buildings in a way that had never been done before. They worked closely from the start with structural engineers to find a way to build their barrier-breaking ideas.

At some point, you have to stop and ask what buildings are for. First and foremost, they have to have an intended purpose: buildings need to function for people. My sense is that all those beautiful buildings of the 1970s and '80s somehow took care of function, but that it wasn't their prime driver. It was really all about the buildings themselves.

Fast forward thirty years to 2000+. Money is not being spent quite so freely as before and clients are looking hard for cost efficiencies and energy costs have also skyrocketed. Architecture and engineering have gone back to the fundamentals. Once again, it's all about integration and collaboration as we work together on designs that will not only make building occupants comfortable, but which will inspire and motivate them. We've come full circle. But now we're not building the same kind of iconic monuments as in the past. Now there is a new appreciation for the environment, an acknowledgment that we have limited natural resources, and a recognition that high-performance buildings are important.

Environmental responsiveness is the major achievement of today's buildings. When I look back at the systems my father designed a half century ago for the Ford Foundation and for the Knights of Columbus headquarters, I realize how incredibly innovative these buildings really were. They were triumphs of building systems integration and sustainability before the concepts formally existed.

Today, a new generation of architects aspires to create great designs, but also to create buildings that will stand the test of time. What is a sustainable building? To me, it's a building that should be there, meaning it hasn't just been plunked down someplace for a short period of time but that it has enduring significance in a specific place, architecturally, culturally, and environmentally. The building defines and enriches that place and fulfills the purpose the building was intended to serve.

The energy crisis that occurred in the mid-to-late 1970s, during the administration of U.S. President Jimmy Carter, sent the country into near panic. What was the response? We reduced air-conditioning, cut down on outside air in buildings, shut off systems, and in general enacted lots of policies and procedures that made building occupants very uncomfortable. We never said, "Wait! Let's make better buildings that use less energy and take advantage of natural ventilation and day-lighting, buildings that make people more comfortable and are environmentally responsive"—until now.

There's been a fundamental change in building design as architects and engineers embrace a thought leadership of much greater consequence than when buildings were primarily driven solely by the economy or corporate identities. Now it's about the big multidisciplinary idea, the holistic response to architectural, social, economic, and environmental challenges. All of this has made the role of the systems engineer much more pivotal and important, so that MEPs are now strategically in the lead instead of being consulted post-design.

There has also been a significant change in expectations. Buildings used to be designed according to the wishes and requirements of owners and occupants as we understood them through our own experiences. But today it is a question of what our children want and need. They live differently than people have ever—plugged in, as they are, to all kinds of constantly evolving electronic information, social media, global connectivity. We therefore have to redefine our approach to buildings and the functions they serve.

All around us there are demands to understand the role of technology, and as a result, there's a more level playing field between architects and systems engineers. Because of the vital significance that integrated building systems play in both the comfort of people and in energy use, systems engineers have assumed a primary role in the design process. Architects now come to us at the very beginning of a project to help create those systems. Very often this has a direct impact on architectural form and siting. What is the best response to the local climate? How should a building be oriented to achieve optimal efficiency? How can we reduce energy consumption? And so on.

Our job is to make buildings come alive with spaces where people can live and work, where they can communicate, and with the systems they need to function efficiently.

Looking ahead, I see the role of the MEP engineer continuing to grow and becoming more and more critical as systems, and increasingly complex systems of systems, define the built environment. Over the past twenty-five years, our services have grown to encompass all the functions people need, including lighting, security, telecommunications, audio/visual, and of course preserving natural resources through sustainable design. I look forward to advancing best practices and innovative solutions while remaining true to the core values that my father has taught by example, not just by words, to me and my colleagues, and on which he firmly based Cosentini for the past sixty years:

1. Listen to the client.
2. Understand what the client wants to accomplish.
3. Never say it can't be done until you have explored all possibilities.
4. Think creatively, and work as a team player to craft new solutions.

I am very fortunate to be able to spend every day of my professional career working with my father and so many wonderful architects. I look forward to continuing the many personal relationships that have developed and to holding high the traditions my father established.

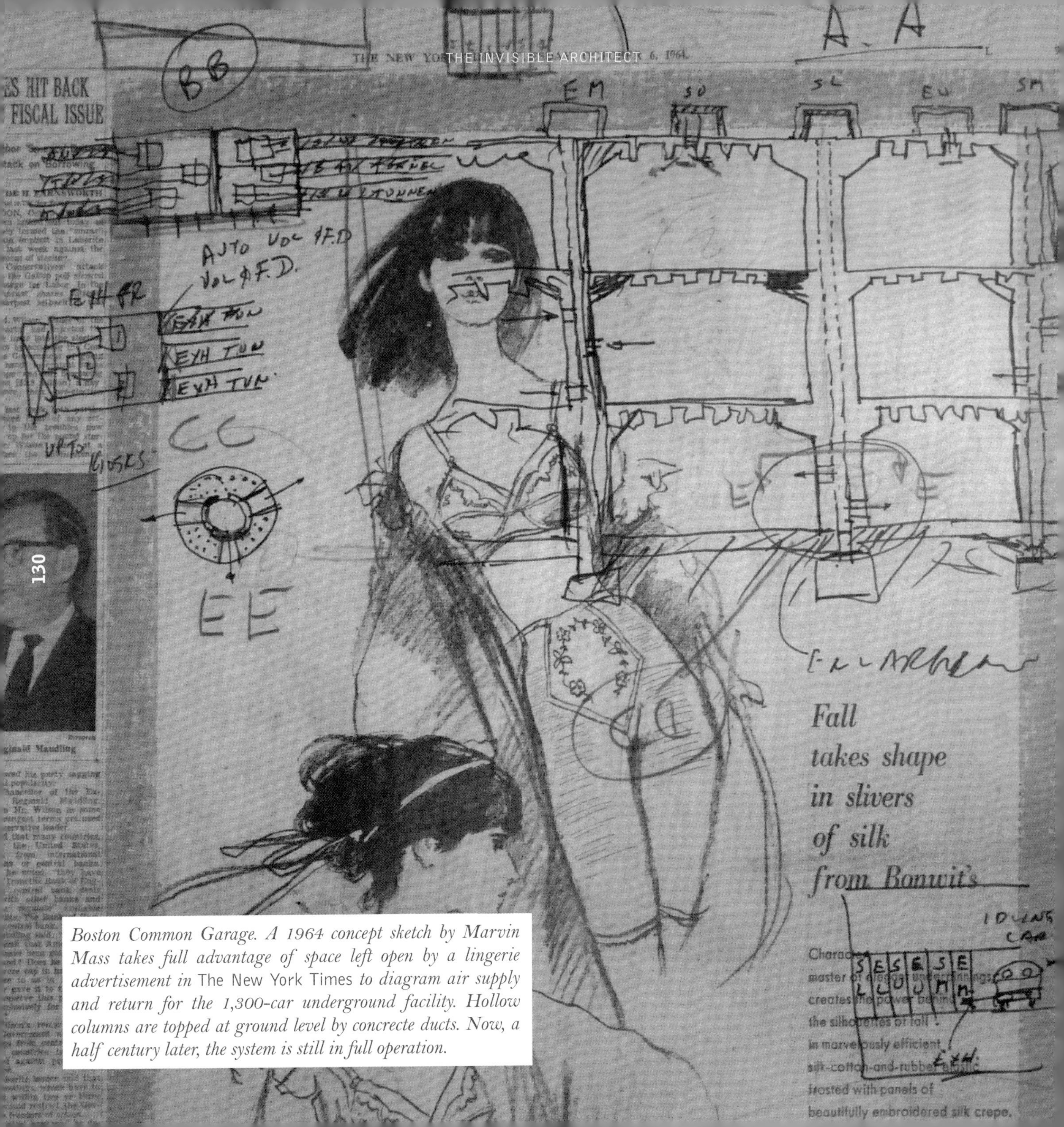

Boston Common Garage. A 1964 concept sketch by Marvin Mass takes full advantage of space left open by a lingerie advertisement in The New York Times *to diagram air supply and return for the 1,300-car underground facility. Hollow columns are topped at ground level by concrecte ducts. Now, a half century later, the system is still in full operation.*

APPENDIX: SELECTED PROJECTS

Over the past 60 years, Cosentini Associates has executed thousands of projects, hundreds of them with the direct personal involvement of Marvin Mass. A selection of projects especially notable for their advances and innovations appears on the following pages.

1950s

PROJECT	ADVANCES AND INNOVATIONS
1. 99 PARK AVENUE NEW YORK, NY (Emery Roth, completed 1954)	Cosentini Job No. 1 Thank you, John Tishman!
2. 460 PARK AVENUE NEW YORK, NY (Emery Roth, completed 1954)	Prefabricated curtain wall building enclosed in one day
3. 666 FIFTH AVENUE NEW YORK, NY (Carson & Lundin, completed 1957)	Another Tishman Family project
4. TIME-LIFE BUILDING NEW YORK, NY (Harrison & Abramovitz, completed 1958)	Our first Sixth Avenue hign-rise

1.

2.

3.

4.

1960s

PROJECT	ADVANCES AND INNOVATIONS
1. **RICHARDS MEDICAL RESEARCH LABORATORIES** UNIVERSITY OF PENNSYLVANIA PHILADELPHIA, PA (Louis Kahn, completed 1960)	In the first known instance, all piping and risers were put outside of the building allowing changes to laboratory interiors without obstruction
2. **BELL LABS HOLMDEL** HOLMDEL, NJ (Eero Saarinen, completed 1964)	First use of reflective glass in building, resulting in innovative mechanical design for reduced heat loads
3. **IBM PAVILION** NEW YORK WORLD'S FAIR (Eero Saarinen Associates, completed 1964)	Air distribution for unusual egg-shaped structure
4. **YOYOGI NATIONAL GYMNASIUM** OLYMPIC STADIUM TOKYO, JAPAN (Kenzo Tange, completed 1964)	Air fountains used for first time and expressed as major design element
5. **CBS HEADQUARTERS** NEW YORK, NY (Eero Saarinen, 1965)	HVAC coordination with building design module—a forerunner of integrated system design
6. **SHRINE OF THE BOOK** JERUSALEM, ISRAEL (Bartos & Keisler, completed 1965)	Designed climate-controlled tube for preservation of the 3,000-year-old Dead Sea Scrolls

1960s

1.

2.

3.

4.

5.

6.

1960s

PROJECT	ADVANCES AND INNOVATIONS
1. HABITAT 67 EXPO 67 WORLD'S FAIR MONTREAL, CANADA (Moshe Safdie, completed 1967)	First prefabricated concrete units for urban housing; units contained individual HVAC systems
2. NEW ENGLAND AQUARIUM BOSTON, MA (Cambridge Seven, completed 1967)	Systems design coordinated for unusual program having public areas, aquarium and lab/work areas
3. FORD FOUNDATION NEW YORK, NY (Roche-Dinkeloo, completed 1968)	Waste air used for air-conditioning in the first major fully enclosed atrium enclosed atrium; innovative energy conservation preceding integrated and sustainable building systems by decades
4. TIME-LIFE BUILDING CHICAGO, IL (Harry Weese, completed 1969)	First use of double-deck elevators to conserve core space
5. KNIGHTS OF COLUMBUS NEW HAVEN, CT (Roche-Dinkeloo, completed 1969)	Integrated system design; ducts and diffusers integrated into architectural elements
6. AETNA HEADQUARTERS HARTFORD, CT (Roche-Dinkeloo, completed 1969)	Integrated systems design; enclosed double beams with holes for HVAC

1960s

1.

2.

4.

5.

3.

6.

1970s

PROJECT	ADVANCES AND INNOVATIONS
1. **GSIS** PHILIPPINES (The Architects' Collaborative/TAC, completed 1970)	Roof flooded for use in evaporative cooling system; early example of green roof engineering resulting in bonus rice crop
2. **STATE UNIVERSITY OF NEW YORK AT ALBANY** ALBANY, NY (Edward Durell Stone, completed 1971)	First district mechanical system for SUNY campus to serve all buildings
3. **SOLAR ONE** UNIVERSITY OF DELAWARE NEWARK, DE (Dr. Mária Telkes, completed 1974)	First house to harvest solar energy in a total-system approach to generate both heat and electricity for domestic use
4. **METROPOLITAN CORRECTIONAL CENTER** NEW YORK, NY (Gruzen Samton, completed 1974)	System design coordinated for separate unit areas
5. **9 WEST 57TH STREET** NEW YORK, NY (Gordon Bunshaft/SOM, completed 1974)	Office building with standard systems adapted to unusually shaped sloped building

1970s

1970s

PROJECT	ADVANCES AND INNOVATIONS
1. **GRAND OPERA HOUSE** GALVESTON, TX (Hardy Holzman Pfeiffer, restoration completed 1974)	Close coordination with acoustical engineer to make sure MEP work did not impair acoustics
2. **JOHN HANCOCK TOWER** BOSTON, MA (Henry N. Cobb/I. M. Pei & Partners, 1976)	First major use of reflective thermopane glass which allowed larger glass walls
3. **FIELD MUSEUM OF NATURAL HISTORY** CHICAGO, IL (Office of Daniel Burnham, 1921; renovated by Harry Weese, 1977)	HVAC for renovated building using original duct shafts
4. **BOETTCHER CONCERT HALL** DENVER, CO (Hardy Holzman Pfeiffer, completed 1978)	Underfloor air return to cover full audience seating
5. **JOHN FITZGERALD KENNEDY PRESIDENTIAL LIBRARY** BOSTON, MA (I. M. Pei/I. M. Pei & Partners, completed 1979)	Diverse, coordinated mechanical systems for glazed atrium, museum spaces, and research center/library
6. **JOHNS MANVILLE HEADQUARTERS** KEN CARYL RANCH, JEFFERSON COUNTY, CO (The Architects' Collaborative/TAC, completed 1979)	Cooling tower used for evaporative cooling to reduce costs

1970s

1.

2.

3.

4.

5.

6.

1980s

PROJECT	ADVANCES AND INNOVATIONS
1. **CRYSTAL CATHEDRAL** GARDEN GROVE, CA (Johnson/Burgee, completed 1980)	Naturally ventilated 86,000-square-foot building; varied mechanical systems result in minimal fuel consumption
2. **500 PARK AVENUE** NEW YORK, NY (James Stewart Polshek, completed 1980)	Coordinated HVAC systems for apartments and office floors in same structure
3. **499 PARK AVENUE** NEW YORK, NY, (James Ingo Freed/I. M. Pei & Partners, completed 1981)	First use of packaged air conditioning units in office tower, including new energy-saving evaporative cooling solutions
4. **HARTFORD SEMINARY** HARTFORD, CT (Richard Meier, completed 1981)	System design for varied spaces and uses
5. **PORTLAND MUNICIPAL SERVICES BUILDING** PORTLAND, OR (Michael Graves, completed 1982)	HVAC systems adapted for public facility program with varied spaces and multiple users
6. **PPG HEADQUARTERS** PITTSBURGH, PA (Johnson/Burgee, completed 1984)	Integrated system designed for multi building complex, ventilation in enclosed winter garden, flood control for basement mechanical floors

1980s

1980s

PROJECT	ADVANCES AND INNOVATIONS
1. **AT&T HEADQUARTERS** NEW YORK, NY (Johnson/Burgee, completed 1984)	Integrated system design for work stations in task oriented program and private offices
2. **McGOVERN INSTITUTE FOR BRAIN RESEARCH** MASSACHUSETTS INSTITUTE OF TECHNOLOGY (MIT) CAMBRIDGE, MA (Goody Clancy, completed 1986)	HVAC design adapted for special use areas with stringent requirements for air filtration, ventilation, etc.
3. **UNITED AIRLINES TERMINAL/ O'HARE INTERNATIONAL AIRPORT** CHICAGO, IL, (Murphy/Jahn, completed 1987)	All ductwork was very carefully integrated with structural and architectural detail
4. **RICE BUILDING** ART INSTITUTE OF CHICAGO CHICAGO, IL (Hammond Beeby Babka, completed 1988)	Renovated and expanded existing plant systems for new museum exhibition space and existing complex
5. **CARNEGIE HALL TOWER** NEW YORK, NY (César Pelli, completed 1991)	First commercial high rise-tower coordinated with Carnegie Hall

1980s

1.

2.

3.

4.

5.

1990s

PROJECT	ADVANCES AND INNOVATIONS
1. **DANIEL PATRICK MOYNIHAN UNITED STATES COURTHOUSE** NEW YORK, NY (Kohn Pedersen Fox, completed 1991)	Complex building with 3 types of circulation, special lighting and acoustics; designed around the judges' requirements
2. **BELL ATLANTIC TOWER** PHILADELPHIA, PA (Kling Lindquist, completed 1991)	60-story tower utilizing factory built and tested package DX units and modular design
3. **STUYVESANT HIGH SCHOOL** BATTERY PARK CITY, NEW YORK, NY (Cooper, Robertson & Partners with Gruzen Samton, completed 1992)	Complex building program including science labs, performance center, sports facilities and academics
4. **FIRST BANK PLACE** MINNEAPOLIS, MN (James Ingo Freed/Pei Cobb Freed & Partners, completed 1992)	Utilized overhead heating in office spaces with extremely cold outside air temperature, radiant heating in lobby
5. **UNITED STATES HOLOCAUST MEMORIAL MUSEUM** WASHINGTON, D.C. (James Ingo Freed/Pei Cobb Freed & Partners, completed 1993)	Special coordination of mechanical devices with architectural details

1990s

1.

2.

3.

4.

5.

1990s

	PROJECT	ADVANCES AND INNOVATIONS
1.	**DISNEY FEATURE ANIMATION BUILDING** BURBANK, CA (Robert A.M. Stern Architects, completed 1994)	First building to house high-tech computer animation equipment to create animated film
2.	**PUERTA DE EUROPA** MADRID, SPAIN (Philip Johnson/John Burgee, completed 1996)	World's first inclined skyscrapers, first high-rise tower in Madrid, still a potent image of Spain's capital
3.	**GUGGENHEIM MUSEUM BILBAO** BILBAO, SPAIN (Frank Gehry, completed 1997)	Integration of MEP with unique architecture and maintain 24/7 climate control for artwork
4.	**MUSEUM OF JEWISH HERITAGE** NEW YORK, NY (Roche-Dinkeloo, completed 1997)	HVAC systems for museum space as part of larger building complex
5.	**RODIN MUSEUM** SEOUL, SOUTH KOREA (Kohn Pedersen Fox, completed 1997)	Ductless A/C system utilizing glass wall cavity for air flow adjustable by season
6.	**4 TIMES SQUARE/CONDÉ NAST BUILDING** NEW YORK, NY (FXFowle, completed 1999)	First green high-rise office building in the United States with PV's, fuel cells, etc.

1990s

1.

2.

3.

4.

5.

6.

PROJECT	ADVANCES AND INNOVATIONS
1. **NEWMAN VERTICAL CAMPUS** BARUCH COLLEGE NEW YORK, NY (Kohn Pedersen Fox, completed 2002)	First vertical university campus in Manhattan
2. **RICHARD B. FISHER CENTER FOR THE PERFORMING ARTS** BARD COLLEGE ANNANDALE-ON-HUDSON, NY (Frank Gehry, completed 2003)	Special acoustical facility to accommodate the varied performing arts requirements
3. **FRANK R. LAUTENBERG SECAUCUS STATION** SECAUCUS JUNCTION, NJ (Brennan Beer Gorman, completed 2003)	Transportation center included track ventilation and smoke control, and infrastructure for future high-rise build-over
4. **WALT DISNEY CONCERT HALL** LOS ANGELES, CA (Frank Gehry, completed 2003)	Total integration of MEP with architecture to achieve acoustical requirements
5. **TIME WARNER CENTER** NEW YORK, NY (David Childs/SOM, completed 2003)	Building includes a multitude of uses ranging from hotel, office, performance venue, retail, etc.
6. **NATIONAL MUSEUM OF THE AMERICAN INDIAN** WASHINGTON, D.C. (Douglas Cardinal, Johnpaul Jones, et al., completed 2004)	24/7 climate control to preserve artifacts and special collections

1.

2.

3.

4.

5.

6.

2000s

PROJECT	ADVANCES AND INNOVATIONS
1. **IAC HEADQUARTERS** NEW YORK, NY (Frank Gehry, completed 2007)	Floor to ceiling walls designed with ceramic frit glass to cut down sun loads so building is shaded in day and glows transparent at night
2. **LEWIS SCIENCE LIBRARY** PRINCETON UNIVERSITY PRINCETON, NJ (Frank Gehry, completed 2009)	Building systems carefully integrated with architecture, updating of systems in existing library and integration of full facility
3. **11 TIMES SQUARE** NEW YORK, NY (FXFowle, completed 2010)	LEED Gold high-rise office building located in Times Square. First to utilize high-efficiency fan-wall technology
4. **NEW WORLD SYMPHONY** MIAMI, FL (Frank Gehry, completed 2011)	Highly varied and customizable performance spaces, distance teaching, and 7,000-square foot exterior video projection wall; designed for maximum flexibility to facilitate traditional and experimental musical performances
5. **SHANGHAI TOWER** SHANGHAI, CHINA (Gensler, est. completion 2014)	121-story tower, second tallest in the world, designed as 8 separate buildings stacked on top of each other, each with its own building systems

1.

2.

3.

4.

5.

vide
lounge
vending
coats
post
kantoor
service point A
service
copy
E-kast
kantoor
RE
RF
RG
RH
RJ

PHOTO CREDITS

Alamy: Nathan Benn: 90 (John Burgee). GlowImages: 120 (top).
Archives of American Art: 67.
Art Institute of Chicago/Ryerson & Burnham Archives: 40 (Weese & Mies), 42, 43, 45 (top).
Artifice Images: Don DeBernardo: 68 (top).
Chicago History Museum: 69 (top), 45 (bottom), 48 (bottom).
Corbis: Richard Cummings: 141. Franz-Marc Frei: 137 (2), 45 (bottom), 48 (bottom).
Cranbrook Archives: 66 (bottom).
ESTO: Ezra Stoller: 36 (top), 55 (top & bottom). Roberto Schezen: 90 (Johnson & Burgee). Luca Vignelli: 90 (Johnson). David Sundberg: 149 (6).
Getty Images/Time & Life Pictures: John Loengard: 37 (top). Frank Scherschel: 47, 49, 53 (left).
University of Arkansas/Fayetteville, Special Collections: 40 (Weese), 54.
University of Pennsylvania: 50, 52, 53 (right).
Yale University of Pennsylvania/Manuscripts and Archives: 40 (Kahn), 58 (left & right), Richard Knight: 59, 62 (section), 64 (all sketches), 66 (top), 68 (bottom), 69, 74 (bottom).

Photos provided courtesy of:

Cosentini Associates: 14, 23, 26, 81, 78 (Pei, Freed), 101, 139 (3), 135 (2), 139 (1), 147 (3). **Araldo Cossutta:** 78 (Cossutta). **Denver Center for the Performing Arts:** Stevie Crecelius: 141 (4); **Michael F. Gebhart:** 141 (6). **Gehry Partners:** 102-103, 105, 110, 153 (4). **Alan Gilbert:** 88, 98, 147 (5). **Frances Halsband:** 92 (bottom). **Hammond Beebe Babka:** 145 (4). **Kohn Pedersen Fox:** Kim Kwan: 149 (5). **Library of Congress:** Carol Highsmith Collection: 121, 141 (6). **National Building Museum:** 72. **Alan Ritchie:** 90 (Alan Ritchie), 98. **Kevin Roche John Dinkeloo and Associates:** 74 (bottom), 76 (top & bottom), 137 (5 & 6). **New York City Landmarks Preservation Commission:** Carl Forster: 71, 74 (top). **Ennead Architects:** 149 (4). **Pei Cobb Freed & Partners:** 36, 78 (Cobb), 83 (section). **Gorchev & Gorchev:** 85, 141 (2). Philip Prouse: 147 (4). C.C. Pei: 82 (top). Frank Rogers: 82 (bottom). Nathaniel Lieberman: 83 (bottom right), 86, 141 (5).

State University of New York/Albany: 57, 139 (2). **Robert A.M. Stern Architects:** 149 (1). **Sitts, George**, "Radio Pierces The Great Blackout", Broadcast Engineering (magazine), December 1965: 61. **Janet Adams Strong:** 11, 33, 32, 34, 44, 46, 48 (top), 51, 56 (top & bottom), 62, 65, 73, 89, 92 (top), 94, 95, 96, 104, 107, 110, 111, 114 (left & right), 115, 116, 117, 119, 122, 124, 126, 130, 133 (1-4), 134 (1), 135 (5), 137 (3, 4), 139 (4, 5), 143 (2), 145 (1), 147 (1), 151 (1), 153 (1-3). **United States Postal Service:** 70 (bottom). Blueprints pages 154 and 157: oldblueprints.com

Wikipedia: Bev Sykes: 70. Philip Capper: 145 (3). Daniel Case: 106, 151(2). Castellana300308: 94 (top). Daderot: 143 (4). Martin Durrschnabel: 145 (5). Jessica Gardner: 149 (2). Tony Hisgett: 109 (top), 149 (3). Hivemind: 135 (3). Derek Jensen: 100, 143 (6). Kanengen: 135 (4). Susan Lesch: 145 (2). Steve Morgan: 118. Andreas Praefcke: 109 (bottom). David Shankbone: 31. Wladyslaw Sojka: 137 (1)

Architects' portraits on page 112 were respectively provided courtesy of:

Max Abramovitz (www.fightingillini.com/coach-law/alumni.html). Edward Larrabee Barnes (photo: Nancy Rica Schiff). Gordon Bunshaft (modarc.wordpress.com/2011/09/06/biography-of-gordon-bunshaft)/. John Burgee. Santiago Calatrava http://reporter.it/archives/6618. Peter Chermayeff. David Childs (photo: Greg Betz). Peter Claman (photo: Steve Friedman). Henry Cobb (photo: Emily Nemens, Center for Architecture). Aldo Cossutta. Joseph Fleischer. Norman Foster (photo: Nigel Young). Bruce Fowle. Robert Fox. James Ingo Freed (Ingbet Gruttner). Frank Gehry (Melissa Majchrzak). Arthur Gensler. Peter Gorman. Michael Graves. Charles Gwathmey (www.paperny.com/gwathmey.html). Hugh Hardy. Walter Hunt. Helmut Jahn. Phillip Johnson (Luca Vignelli/ESTO). Gene Kohn. Morris Lapidus. (www.domusweb.it). Jill Lerner. Ralph Mancini. Richard Meier. William Pedersen. I.M. Pei (Ingbet Gruttner). Cesar Pelli (Peter Hurley). James Stewart Polshek. Alan Ritchie. Kevin Roche. Nancy Ruddy. Peter Samton. Robert A.M. Stern. Marilyn Jordan Taylor.

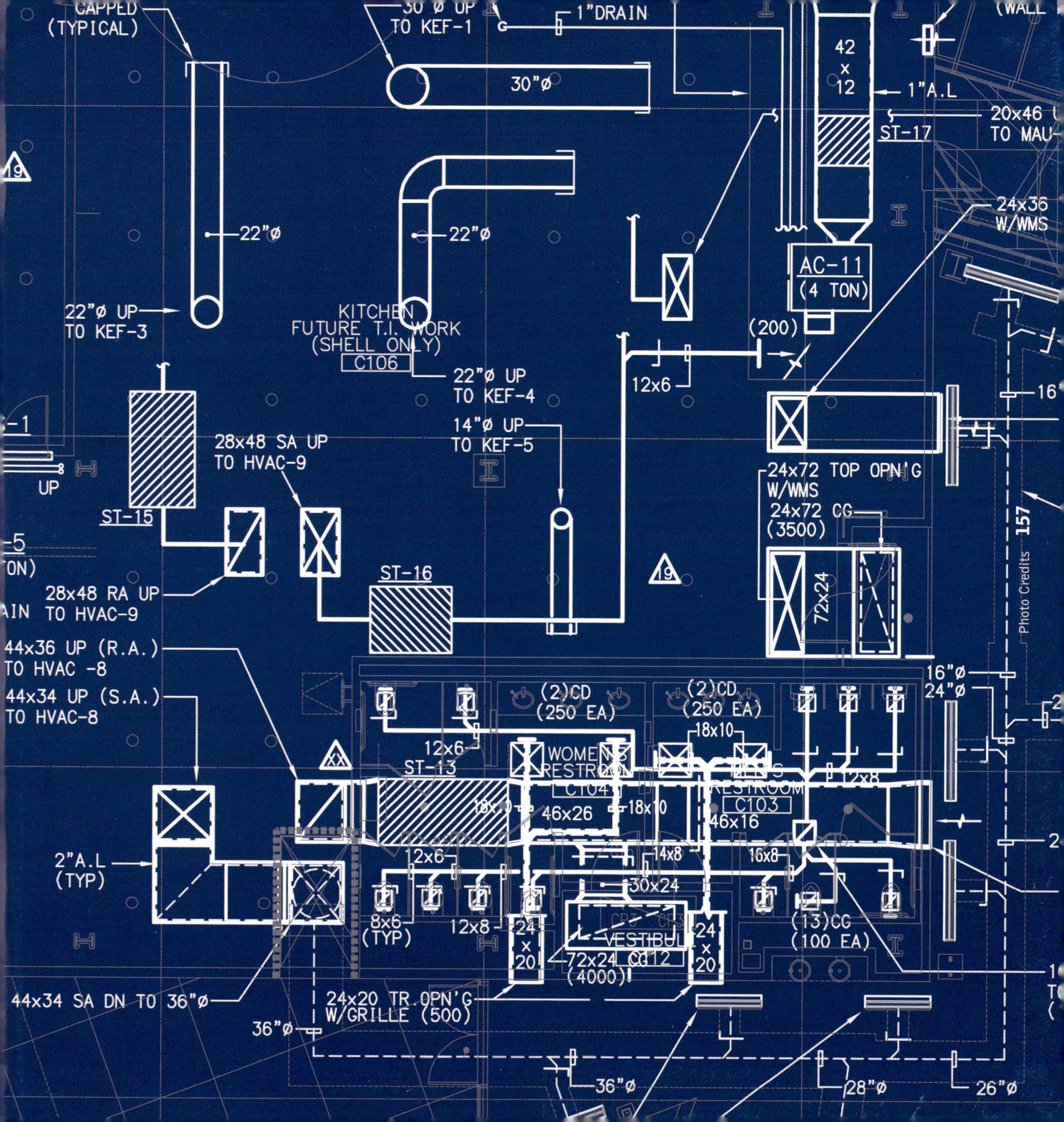

CAPPED
(TYPICAL)
30"ø UP
TO KEF-1
1"DRAIN
30"ø
42
x
12
1"A.L
20x46
TO MAU
ST-17
22"ø
22"ø
AC-11
(4 TON)
24x36
W/WMS
22"ø UP
TO KEF-3
KITCHEN
FUTURE T.I. WORK
(SHELL ONLY)
C106
22"ø UP
TO KEF-4
(200)
12x6
14"ø UP
TO KEF-5
28x48 SA UP
TO HVAC-9
UP
ST-15
24x72 TOP OPN'G
W/WMS
24x72 CG
(3500)
72x24
157
28x48 RA UP
TO HVAC-9
ST-16
44x36 UP (R.A.)
TO HVAC -8
44x34 UP (S.A.)
TO HVAC-8
(2)CD
(250 EA)
(2)CD
(250 EA)
16"ø
24"ø
18x10
12x6
ST-13
WOMENS
RESTROOM
C104
RESTROOM
C103
12x8
18x10
46x26
18x10
46x16
2"A.L
(TYP)
12x6
14x8
16x8
30x24
8x6
(TYP)
12x8
24
x
20
VESTIBULE
C012
24
x
20
(13)CG
(100 EA)
72x24 CG
(4000)
44x34 SA DN TO 36"ø
24x20 TR.OPN'G
W/GRILLE (500)
36"ø
36"ø
28"ø
26"ø
Photo Credits

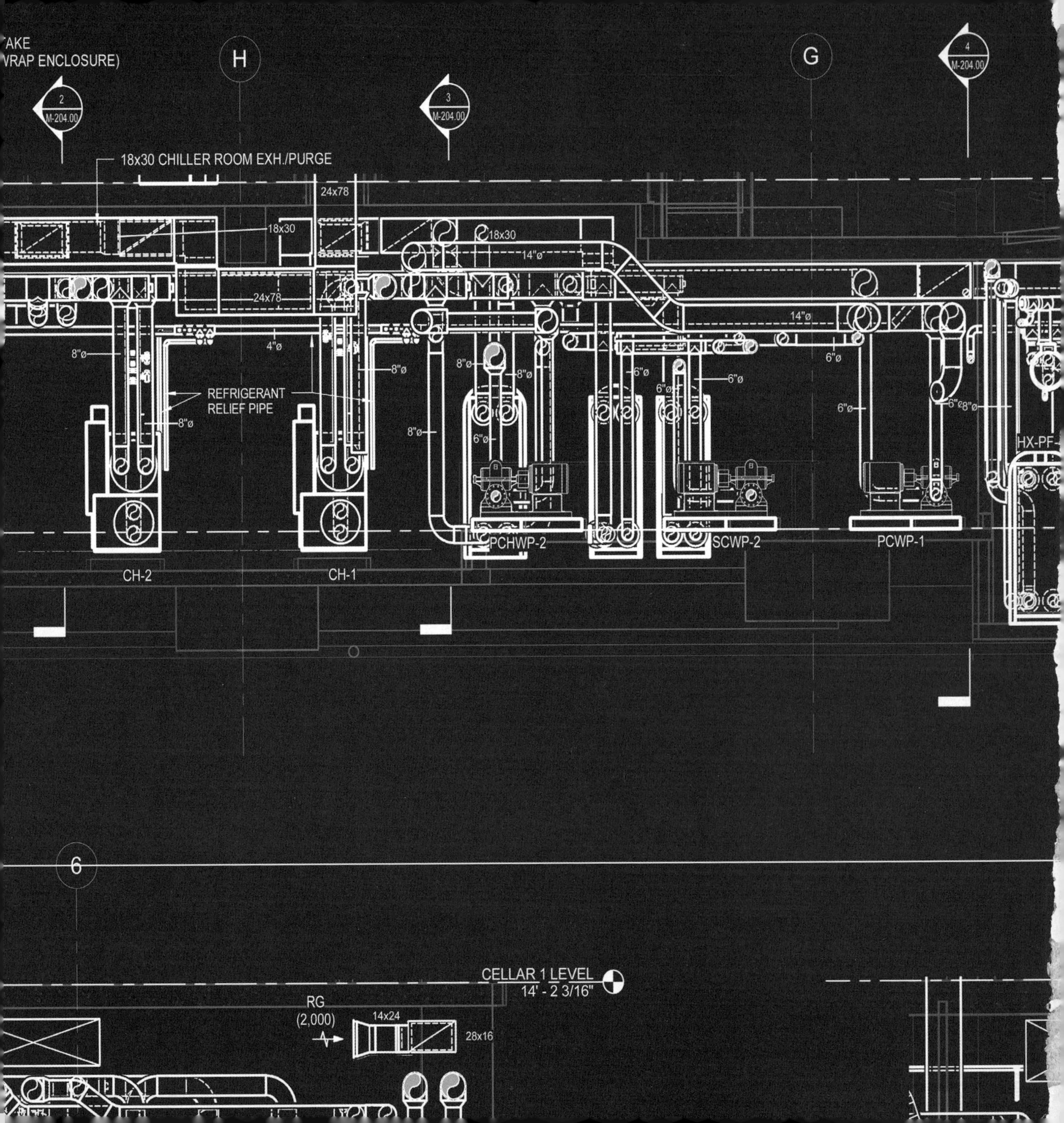

'AKE
VRAP ENCLOSURE)
H
G
2
M-204.00
3
M-204.00
4
M-204.00
18x30 CHILLER ROOM EXH./PURGE
24x78
18x30
18x30
14"ø
24x78
14"ø
4"ø
8"ø
8"ø
8"ø
8"ø
6"ø
6"ø
6"ø
6"ø
6"ø
6"ø
8"ø
REFRIGERANT
RELIEF PIPE
8"ø
8"ø
6"ø
HX-PF
PCHWP-2
SCWP-2
PCWP-1
CH-2
CH-1
6
CELLAR 1 LEVEL
14' - 2 3/16"
RG
(2,000)
14x24
28x16

F
5
M-204.00
E
6
M-204.00
CELLAR 1 LEVEL
14' - 2 3/16"
30"BOILER FLUE
30"BOILER FLUE
18x30
2"FOS&R
(PIPES IN PIPE)
12"LPS HEADER
12"LPS HEADER
STEAM TRAP
2 1/2"ø
1 1/2"ø
20ø
2 1/2"ø
2 1/2"ø
1 1/2"ø
2 1/2"ø
B-1
B-2
B-3
BFT-1
BDS
CELLAR 2 LEVEL
-1' - 0 13/16"
1 1/2"ø
1 1/2"ø
TO F.D.
TO F.D.
2 1/2"MAKE-UP
1
M-204.00
6
CELLAR 1 LEVEL
14' - 2 3/16"
CHWS
CHWR

continued from front inside cover

ARCHITECTS WHO HAVE WORKED WITH COSENTINI ASSOCIATES, 1952 - 2012

• KMD Architects • Knoll Roslyn Design Group • Koenen Associates • Koetter/Kim & Associates • Kohn Pedersen Fox Associates • Kohn Pedersen Fox Conway • Kohnke Architect • Costas Kondylis & Partners • Kondylis Design • Costas Kondylis & Associates • Rem Koolhaas • Kossar + Garry Architects • Kostow Greenwood Architects • Koulbanis Brandreth Associates • Koutsomitis Architects • Alexander Kouzmanoff Architect • Edward Kozanlian • KPA Design Group, Inc. • KPF Interior Architects • KPF Associates • Benjamin D. Kracauer Architect • Krueck & Olsen Architects • Howard Kulp Architects • George Kunihiro Architect & Associates • Kunwon Architects • Kisho Kurokawa Architect & Associates • Kutnicki Bernstein Architects • Kyu Sung Woo, Architects • **L** • Lambert Woods Architects • Landy Verderame Arianna • Alan Lapidus Architects • Morris Lapidus • Morris Lapidus & Associates • Large-Moger & Associates • Larsen Associates Architecture • Larsen Shein Ginsberg Snyder • LCG Architects • Steven J. Leach & Associates • Lee Harris Pomeroy Architects • Leers Weinzapfel Associates • Thomas Leeser Architects • William Leggio Architect • Leonard, Colangelo & Peters • Stephen Lepp Associates• William Lescaze • Jonathan Levi Architects • Levitt & Sons • R. Lewin Interior Design • Charles Lewis • Neville Lewis Associates • Ismael Leyva Architects • David Liametz Associates • Studio Daniel Libeskind • Liebman Associates • Liederbach & Graham, Architects • Lindsay Newman Architecture & Design • Linde-Hubbard • Mark D. Lipton Architect • LiRo Architects & Planners • Llewelyn Davies & Associates • Loffredo Brooks Architects • Lohan Associates • Lohan Caprile Goettsch Architects • Joseph Pell Lombardi & Associates • John LoPinto • Lothrop Associates • James D. Lothrop • Luce et Studio • Lucien Lagrange Architects• Charles Luckman Architect • Oliver Lundquist & Associates • Victory Lundy • LWC Design • Robert G. Lyon & Associates (RGLA) • Lyras, Galvin and Anaya • **M** • Machado and Silvetti Associates • Michel Macary • Machinist Associates Architects • MACK Architects • J. P. Maggio Design Associates • Maki and Associates • Avinash K. Malhotra Architects • Mancini-Duffy • Herbert Mandel • Margolis + Fishman • Peter Marino Architect • Mark Mariscal Architects • Mascioni & Behrman • Massa Montalto Architects • May & Pinska • Mayer Whittlesey & Glass • Mayers & Schiff Associates • McClier Architects and Engineers • McDonald Associates • McDonald Becket • Willliam McDonough + Partners • McDonough Rainey Architects • Richard McElhiney Architects • MCG Architects • John J. McNamara • McMillan Inc. • MDM Design Group • William Meek Associates • Richard Meier & Partners • Meli/Borelli Associates • Stephen Melnick • Meltzer/Mandl Architects • John R. Menz Architects • David K. Mesbur • Metcalf Tobey • Metcalf Tobey Davis • George Metzger • MG Design International • MGA Partners Architects • Miceli Kulik Williams & Associates • Office of Mies van der Rohe • Millard Schroeder • Sidney Miller • Miller Savarese Associates • Mitchell/Giurgola Architects • MKDA • MKW & Associates • Moed de Armas & Shannon • Mojo Stumer Associates • Molyneux Architects • Montroy Andersen • Montroy Andersen Design Group • Montroy Andersen DeMarco • Morgan Wheelock • Toshiko Mori Architect • Joseph Morog Architect • S. I. Morris • Morris+Aubrey Architects • Morrison Murakami Architects • Mosscrop Associates • Myron Moss • Ted Moudis & Associates • Moyer Associates • MS Architecture • Muchow Associates • Mullen Palandrani Architects • C.F. Murphy Associates • Murphy/Jahn • **N** • Nadel Architects • Nagle Hartray Architecture • Naiztat + Ham • J. T. Nakaoka Associates • Nasr, Penton & Associates • Gal Nauer Architects • NBBJ • Jonathan Nehmer + Associates, Inc. • Nelson Interior Design • Neuham Taylor & Bernard Rothzeid • New York Design Collaborative • Herbert S. Newman • Nichols Brosch Sandoval & Associates • NJK-12 Architects • Notter Finegold + Alexander • Eliot Noyes & Associates • **O** • O'Donnell Wicklund Pigozzi & Peterson • Office Design Associates • Office for Metropolitan Architecture (OMA) • Oger International • H. Thomas O'Hara • O'Hare Associates • Oldham & Partners • Oldham & Seltz • Olin Partnership • Olson Lewis + Architects • Olson Group • Onyx Architects • Oppenheimer, Brady & Lehrecke • Oppenheimer Brady & Vogelstein • Carlos A. Ott • Owen & Mandolfo • OWP/P • **P** • Pagnamenta Torriani • Pallante Design Associates • Steven P. Papadatos • Papadatos & Moudis Associates • Pappageorge/Haymes • John Austin Parker • Leonard Parker Associates • Pasanella + Klein Stoltzman + Berg Architects • Charles Patten, AIA • Jillian Paul Interiors • Peckham Guyton Albers & Viets • PEG/Park • I. M. Pei & Partners • Pei Cobb Freed & Partners • Pei Partnership Architects • Leo C. Peiffer & Associates • César Pelli & Associates • Pelli Clarke Pelli Architects • Pembrooke & Ives • Pereira & Luckman • Perfido Weiskopf • Bradford Perkins Architect • Perkins Eastman Architects • Perkins Eastman Larsen Shein • Perkins & Will • Perkins, Will/Russo & Sonder • Perkins & Will Partnership • Perry, Dean & Stewart • Peterson & Brickbauer • Peterson Griffen Architects • Thomas Phifer & Partners • The Phillips Group (TPG) • Renzo Piano Building Workshop • Pickard Chilton Architects • Alberto Pinto • Planned Expansion Group • Planned Space Interiors, Inc. • Platt Byard Dovell • Platt Wyckoff & Coles • Plaza • Jan Hird Pokorny • Polatnick Zacharjasz Architects • James Stewart Polshek Architect • Polshek Partnership • Pomerance & Breines • Alfred Easton Poor • Atelier Christian de Portzamparc • Jennifer Post Design • Powers & Kessler • Pratt Box & Henderson • Preiss Breismeister Architects • Prellwitz/Chilinski Associates • Prentice, Chan, Ohlhausen • The Presnick Group • Proposition Architects • Pruyn-Bergren Associates • PS&S Architecture • **R** • Rabun Hatch & Associates • Rabun Hogan Ota Rasche Architects • Rick Rakusin • RAL Companies • Janko Rasic Architects • William Rawn Associates Architects • Raymond & Rado • Raymond, Rado, Caddy & Bonington • The RBA Group • RBSD Architects